Springer Series in Statistics

Advisors:
P. Bickel, P. Diggle, S. Fienberg, U. Gather,
I. Olkin, S. Zeger

Springer Series in Statistics

Alho/Spencer: Statistical Demography and Forecasting.
Andersen/Borgan/Gill/Keiding: Statistical Models Based on Counting Processes.
Atkinson/Riani: Robust Diagnostic Regression Analysis.
Atkinson/Riani/Cerioli: Exploring Multivariate Data with the Forward Search.
Berger: Statistical Decision Theory and Bayesian Analysis, 2nd edition.
Borg/Groenen: Modern Multidimensional Scaling: Theory and Applications, 2nd edition.
Brockwell/Davis: Time Series: Theory and Methods, 2nd edition.
Bucklew: Introduction to Rare Event Simulation.
Cappé/Moulines/Rydén: Inference in Hidden Markov Models.
Chan/Tong: Chaos: A Statistical Perspective.
Chen/Shao/Ibrahim: Monte Carlo Methods in Bayesian Computation.
Coles: An Introduction to Statistical Modeling of Extreme Values.
Devroye/Lugosi: Combinatorial Methods in Density Estimation.
Diggle/Ribeiro: Model-based Geostatistics.
Dudoit/Van der Laan: Multiple Testing Procedures with Applications to Genomics.
Efromovich: Nonparametric Curve Estimation: Methods, Theory, and Applications.
Eggermont/LaRiccia: Maximum Penalized Likelihood Estimation, Volume I: Density Estimation.
Fahrmeir/Tutz: Multivariate Statistical Modelling Based on Generalized Linear Models, 2nd edition.
Fan/Yao: Nonlinear Time Series: Nonparametric and Parametric Methods.
Ferraty/Vieu: Nonparametric Functional Data Analysis: Theory and Practice.
Fienberg/Hoaglin: Selected Papers of Frederick Mosteller.
Frühwirth-Schnatter: Finite Mixture and Markov Switching Models.
Ghosh/Ramamoorthi: Bayesian Nonparametrics.
Glaz/Naus/Wallenstein: Scan Statistics.
Good: Permutation Tests: Parametric and Bootstrap Tests of Hypotheses, 3rd edition.
Gouriéroux: ARCH Models and Financial Applications.
Gu: Smoothing Spline ANOVA Models.
Györfi/Kohler/Krzyżak/Walk: A Distribution-Free Theory of Nonparametric Regression.
Haberman: Advanced Statistics, Volume I: Description of Populations.
Hall: The Bootstrap and Edgeworth Expansion.
Härdle: Smoothing Techniques: With Implementation in S.
Harrell: Regression Modeling Strategies: With Applications to Linear Models, Logistic Regression, and Survival Analysis.
Hart: Nonparametric Smoothing and Lack-of-Fit Tests.
Hastie/Tibshirani/Friedman: The Elements of Statistical Learning: Data Mining, Inference, and Prediction.
Hedayat/Sloane/Stufken: Orthogonal Arrays: Theory and Applications.
Heyde: Quasi-Likelihood and its Application: A General Approach to Optimal Parameter Estimation.
Huet/Bouvier/Poursat/Jolivet: Statistical Tools for Nonlinear Regression: A Practical Guide with S-PLUS and R Examples, 2nd edition.
Ibrahim/Chen/Sinha: Bayesian Survival Analysis.
Jiang: Linear and Generalized Linear Mixed Models and Their Appliations.
Jolliffe: Principal Component Analysis, 2nd edition.
Knottnerus: Sample Survey Theory: Some Pythagorean Perspectives.

(continued after index)

Pierre Lavallée

Indirect Sampling

Springer

Pierre Lavallée
Statistics Canada
100 Tunney's Pasture Driveway
15th Floor, R.H. Coats Bldg.
Ottawa, Ontario
K1A OT6 Canada
pierre.lavallee@statcan.ca

ISBN 978-1-4419-2421-6 e-ISBN 978-0-387-70782-2
DOI 10.1007/978-0-387-70782-2

Printed on acid-free paper.

9 8 7 6 5 4 3 2 1

springer.com

"Common sense is the most equally shared thing in the world."

DESCARTES

FOREWORD

"Writing a foreword is a formidable honour." These words come from one of my friends who, in 1988, began in this manner his preface that he had kindly written for one of my books. It is only today that I truly realise the complete accuracy of his sentiments.

It is without a doubt an honour, and I definitely feel this way about it, for this is an excellent work and its author is strongly captivating.

A mathematician who graduated with highest honours from the University of Ottawa, Pierre Lavallée conquered the lofty goal of a Masters in Science (mathematics and statistics option) at Carleton University. For more than fifteen years, he has held the position of senior survey methodologist at Statistics Canada, where he could supplement his existing theoretical training with solid experience in one of the most outstanding official organisations in the field of surveys. It was therefore with a great deal of enthusiasm that I supervised his doctoral thesis that he brilliantly defended in June 2001 at the *Université Libre de Bruxelles* and from which this book evolved.

During the second half of the 20th century, we saw more and more books on survey theory, a movement that continues at the start of this new millennium. Many of them are good, and even very good. This is the case, for example, with the book written by Carl-Erik Särndal, Bengt Swensson and Jan Wretman (1992) that some consider — a justifiable title by the way — as the reference book of the end of the 20th century for all scientists working in this domain. How, under these conditions, can we still propose a written work that keeps this attention of the public if it does not reach new means, expanding the facets that we generally find in these works, granting the privilege of an educational presentation from a book?

In a text entitled *« Dans quelle direction vont la théorie et la pratique des sondages ? »* ("In which direction are the theory and practice of sampling headed?"), which the reader can refer to in the book that I edited for Dunod with Ludovic Lebart in 2001 (p. 20), Carl-Erik Särndal insists on the fact that in scientific literature, "survey methodology and sampling theory are certainly two different things." Very few people can boast that they possess the recognised competence in the domains of both *sampling theory* and *survey practice*. Pierre Lavallée is part of a small, fortunate group that concurrently holds these qualities and who, in fact, can only enrich the disciplines in which he works.

Furthermore, the discourse is new. Up to now, the majority of proposed sampling methods looked to estimate parameters of a population by taking a sample selected directly from a sampling frame consisting of units from that population. The idea defended by Pierre Lavallée goes further since it proposes to estimate these parameters by sampling not the population concerned, but another population having connections with the first one. Look for information about children by selecting parents or obtain information on subsidiaries of companies through the parent companies, all while conserving the statistical properties of the estimators so constructed; these depict examples of actual concrete problems for which the proposed approach offers elegant solutions. We must be indebted to Pierre Lavallée for having detailed this issue and for presenting it to us with the pedagogical qualities that are so familiar to us all.

Before concluding, I asked myself if this preface sufficiently and clearly conveyed all of the goodness that I thought of this book and of its author if the reader read it from beginning to end, but I am half reassured by calling to mind what was said by Luc de Clapiers, marquis de Vauvenargues, in his thoughts: "I have never seen a boring preface leading into a good book."

May this work have all the success that it merits.

Jean-Jacques Droesbeke
Université Libre de Bruxelles
April 2002

PREFACE

Among all books written on sampling theory, there was no existing one devoted to *indirect sampling*. In 2002, I published in French the book "*Le sondage indirect, ou la méthode généralisée du partage des poids*" at the *Éditions de l'Université de Bruxelles* (Belgium) and the *Éditions Ellipse* (France). The present book on *indirect sampling* is a translated version of this book, with some sections added to reflect the new developments that have occurred since 2002. For the readers that are familiar with the content of the previous book, the new developments are with respect to obtaining optimal weighted links (section 4.6.3), the treatment of the problem of links identification (section 8.7), and some recent applications (chapter 10).

As we know, sampling may be performed by drawing samples of people, businesses, or other things that we survey in order to obtain the desired information. According to classical sampling theory, the selection of samples is done by selecting at random from lists called sampling frames. These sampling frames are supposed to represent the set of people or businesses for which we are looking to produce information; this is what constitutes the target population.

When the statistician has a sampling frame representing the desired target population, the drawing of samples can be made according to the well-established techniques of classical sampling theory, which we could also call direct sampling, as opposed to *indirect sampling*. The techniques of classical theory depend on a random selection of samples so that we can establish the probability of drawing some sample or other. This is what we call probability sampling. The knowledge of the selection probability ensures that we can establish the precision and reliability of the information produced by the survey. For example, we can establish if the results produced can contain a bias and in which interval we can hope to find the "true" response. Moreover, the selection probabilities are directly used in the calculations of the results in obtaining precise estimates.

In certain situations, the survey statistician does not have at his or her disposal any sampling frame and he or she must then manage to construct the samples needed in order to obtain the desired information. For opinion polls, for example, it is not rare that the sample of respondents be obtained by surveying the opinion of people chosen at random in a shopping centre. Since we do not have a list of customers at the shopping centre, there is then an absence of a sampling frame. This absence ensures that we cannot establish the probability of obtaining the sample, which makes the calculation of sampling precision impossible. This type of sampling is described as non-probability sampling.

In other situations, the survey statistician has access to sampling frames, but none of which correspond directly to the desired target population. To carry out the survey, the statistician can then choose a sampling frame that is indirectly related to it. For example, for a survey about children, the statistician can make use of a sampling frame of adults whose children are chosen to be surveyed. In this case, the statistician first selects a sample of adults from the sampling frame which he or she has. For each adult in the sample, the statistician then identifies the children of the selected adult and finally surveys on behalf of all of the children identified. This is in the end what we mean by *indirect sampling*. Let us note that since there is the usage of a sampling frame, indirect sampling is a form of probability sampling.

Indirect sampling finds its application in social surveys, as seen previously, but also in economic surveys. For example, for a survey about businesses, the survey statistician can consider the possibility of using, by way of a sampling frame, a list of businessmen or businesswomen registered at a chamber of commerce. Indirect sampling becomes complicated here when a businessperson owns more than one business, or when a business is the property of more than one businessperson.

One sampling technique that can be as often employed in the context of a classical sampling, as it is in *indirect sampling,* is cluster sampling. This sampling technique is not used for the sample selection of units (people, businesses, or others), but instead for samples of groups of units called clusters. In social surveys, clusters most often correspond to households or dwellings. In fact, a dwelling consists of a cluster of persons living in it. In economic surveys, clusters are generally enterprises that own establishments.

When *indirect sampling* is used jointly with cluster sampling, many complications stand out for the survey statistician. One of these complications lies in the calculation of the selection probabilities of surveyed units at the time of the survey. As was mentioned previously, the knowledge of the selection probabilities allows for the establishment of the precision and reliability of the information produced by the survey. Furthermore, these are directly used in the calculation of the results derived from the survey. The knowledge of the selection probabilities is therefore considered as vital for the survey statistician.

In the absence of selection probabilities, it is possible to calculate values that can substitute for the selection probabilities and can produce survey results that are entirely as valid for the survey statistician as for users of the results (governments, company directors, sociologists, etc.). This is possible under the *generalised weight share method (GWSM).* In sampling theory, weights are generally associated with the inverse of the selection probabilities. The *GWSM* in part uses the selection probabilities in a relatively simple calculation focused on the relationship between units from the sampling frame and those from the target population. In the context of *indirect sampling*, let us recall that the sampling frame and the target population are distinct.

The use of the *GWSM* proves to be crucial in the context of *indirect sampling*, and in particular in the *indirect sampling* of clusters. The production of estimates of simple totals or means can often become almost insurmountable without this method. The *GWSM* in fact allows for the solution of problems, both theoretical and practical, that up to now gave nightmares to survey statisticians.

The development of *indirect sampling* and the *GWSM* is the fruit of many years of reflection from the solution of practical problems occurring in the application of classical sampling theory. The lack of a sampling frame for a target population unfortunately constitutes a very common situation, even in national statistical institutes. This is what brought me to the publication of this book. I hope that survey statisticians will find in it answers to their questions and that they will be able to put into practice the different developments presented about *indirect sampling* and the *GWSM*.

In closing, I would like to greatly thank Jean-Jacques Droesbeke of the *Université Libre de Bruxelles* (ULB), who patiently encouraged me to write this book, to whom I also express my gratitude. I would also like to thank Jean-Claude Deville of the *École Nationale de la Statistique et de l'Analyse de l'Information* (ENSAI) for his invaluable advice and especially for the spark behind the generalisation of the weight share method. My thanks go also to Carl Särndal, who pertinently gave me advice in the formulation of certain theoretical results. My gratitude also goes to my colleague Pierre Caron of Statistics Canada, who carried out the simulations, and to my other colleague Michel Hidiroglou of Statistics Canada, who is a daily source of inspiration. Also, I would like to thank the late Bernard Gailly of the *Centre d'Études des Populations, de la Pauvreté et des Politiques Socio-économiques* (CEPS) in Luxembourg, with whom the numerous discussions were always fruitful. I should surely not forget Pambu Kita-Phambu from the ULB who helped in the layout of this book, and also my colleague Leon Jang who translated it from French to English. Finally, I wish to thank my wife Marie-Claude, who encouraged and supported me in the writing of this book.

TABLE OF CONTENTS

Chapter 1: Introduction 1

1.1 Review of sampling theory and weighting 1

1.2 Cluster sampling 4

1.3 Indirect sampling 7

1.4 Generalised weight share method 10

Chapter 2: Description and Use of the GWSM 13

2.1 Description 13

2.2 Use 18

2.2.1 Indirect sampling for rare populations 19

2.2.2 Weighting using only the selection probabilities of the selected units 20

2.2.3 Weighting of populations related by complex links 20

2.2.4 Weighting of unlinked units 21

Chapter 3: Literature Review 23

3.1 First steps 23

3.2 Fair Share Method 24

3.3 Contribution of Ernst (1989) 28

3.4 Network Sampling 32

3.5 Adaptive Cluster Sampling 37

3.6 Snowball Sampling 42

Chapter 4: Properties 45

4.1 Bias and variance 45

4.2 Particular case 1: cluster sampling 50

4.3 Particular case 2: census of population U^A 53

4.4 Particular case 3: census of population U^B 54

4.5 Use of weighted links 56

4.6 Improvement of the estimator 59

4.6.1 Conditional approach 59

4.6.2 Use of sufficient statistics 63

4.6.3 Obtaining optimal weighted links 65

Chapter 5: Other Generalisations 77

5.1 Two-stage indirect sampling 77

5.2 Arbitrary aspect in the formation of clusters 83

5.2.1 Extreme case (i): population U^B with a single cluster of size M^B 84

5.2.2 Extreme case (ii): population U^B with M^B clusters of size 1 86

5.2.3 General case and discussion 89

5.3 Elimination of the notion of clusters 97

Chapter 6: Application in Longitudinal Surveys 105

6.1 Sampling design of SLID 106

6.1.1 Initial sample 107

6.1.2 Supplementary sample ... 109

6.2 Estimation weights ... 111

6.3 Use of the GWSM in obtaining estimation weights ... 113

6.4 Variance estimation ... 117

6.5 Use of another type of links ... 118

Chapter 7: GWSM and Calibration ... 121

7.1 Review of calibration ... 121

7.2 GWSM with calibration ... 128

7.3 Particular case 1: auxiliary variables coming from U^A ... 132

7.4 Particular case 2: auxiliary variables coming from U^B ... 136

7.4.1 Application of calibration before GWSM ... 136

7.4.2 Application of calibration after GWSM ... 138

7.4.3 Comparison of the two approaches ... 142

7.4.4 Simulation study ... 145

Chapter 8: Non-response ... 151

8.1 Types of non-response ... 152

8.2 Correcting response rates ... 155

8.3 Response probabilities ... 157

8.4 Treatment of non-response within s^A ... 160

8.5 Treatment of cluster non-response ... 166

8.6 Treatment of unit non-response ... 174

8.7 Treatment of errors in links identification ... 183

8.7.1 Record linkage ... 186

8.7.2 Modelling ... 187

8.7.3 Estimating the proportion of links 189

8.7.4 Calibration 190

8.7.5 Proportional adjustments 190

Chapter 9: GWSM and Record Linkage 195

9.1 Record linkage 196

9.2 GWSM associated with record linkage 199

9.2.1 Approach 1: use all non-zero links with their respective linkage weights 201

9.2.2 Approach 2: use all non-zero links above a given threshold 204

9.2.3 Approach 3: choose the links randomly 206

9.2.4 Some remarks 210

9.3 Simulation study 211

9.3.1 Data used 212

9.3.2 Sampling plan 214

9.3.3 Results and discussion 215

Chapter 10: Conclusion 225

Notations 231

Bibliography 235

Index 243

CHAPTER 1

INTRODUCTION

Sample surveys today make up a varied and often indispensable source of information. Whether at the level of governments, company managers, sociologists, economists, or ordinary citizens, surveys allow the informational needs necessary in taking a decision to be met. For example, to establish their policies concerning certain economic sectors, governments must have a picture of the situation before taking decisions concerning these sectors.

1.1 REVIEW OF SAMPLING THEORY AND WEIGHTING

Sample surveys are carried out by selecting samples of persons, businesses or other items (called *units*) that we survey in order to get the desired information. Sample selection is often done by randomly selecting certain units from a list that we call a *sampling frame*. This list, or sampling frame, is supposed to represent the set of units for which we are looking to produce information; this is what makes up the *target population*. The sample size can be determined prior to the selection (*fixed size sampling*) or at the time of the sampling itself (*random size sampling*). In this book, we will restrict ourselves to fixed size sampling which is, in practice, the most widespread.

Strictly speaking, fixed size sampling is described as follows. Consider $\mathbf{Y}_U = (y_1,...,y_N)$, the vector containing the values y_k for a population U of size N. For a survey on tobacco use, for example, the variable of interest y_k of $\mathbf{Y}_U$ can be the number of cigarettes smoked by individual k during a given day. In general, we want to know the value for instance of the total $Y = \sum_{k=1}^{N} y_k$, or otherwise the mean $\overline{Y} = Y / N$. If the size N of population U is known, the problem in determining the total Y or the mean $\bar{Y}$ is the same. Going back to the previous example on tobacco, the total Y represents the total number of cigarettes smoked during the day, while the mean $\bar{Y}$ represents the average number of cigarettes smoked by an individual.

To estimate the total Y (or the mean $\bar{Y}$) of population U, we select a sample s of size n. A *sampling design* $\mathbf{p}$ is a function $\mathbf{p}(s)$ of the set Ξ of all samples s selected from U such that $\mathbf{p}(s) \geq 0$ and $\sum_{s \in \Xi} \mathbf{p}(s) = 1$. The function $\mathbf{p}(s)$ is in fact the probability of selecting sample s among all samples of Ξ. We assume that $\mathbf{p}(s)$ is known for the set Ξ ; this is what we call *probability sampling*. A well-known sampling design is *simple random sampling* (without replacement) where all possible samples of Ξ have the same chance of being selected. We have in fact $\mathbf{p}(s) = n!(N-n)!/N!$. By dividing the population U into subpopulations U_h called *strata*, where $U = \bigcup_h U_h$, we define *stratified simple random sampling* that consists of selecting a simple random sample in each of the h strata.

We define the *selection probability* (or *inclusion probability*) of unit k from population U by

$$\pi_k = \sum_{s \ni k} \mathbf{p}(s), \tag{1.1}$$

where the sum of (1.1) is carried out over all the samples of s from the set Ξ that contains unit k. We assume that $\pi_k > 0$ for all units k of population U, i.e., all units have a non-zero chance of being selected. For example, with simple random sampling, we get $\pi_k = n/N$, for $k=1,...,N$.

For each unit k of s, we measure the value of the variable of interest y_k. We can estimate the total Y with the following *Horvitz-Thompson estimator*:

$$\hat{Y}^{HT} = \sum_{k=1}^{n} \frac{y_k}{\pi_k}, \tag{1.2}$$

where the sum of (1.2) is carried out over all units k of sample s (Horvitz and Thompson, 1952).[1] We can show that the estimator $\hat{Y}^{HT}$ is

[1] In this book, the sums will be based on a re-indexing of units. For example, for a sum over the population of size N and another over the sample of size n selected

unbiased for Y with respect to the sampling design, i.e., that if $\hat{Y}_s^{HT}$ represents the value of $\hat{Y}^{HT}$ obtained for sample s, we have:

$$E(\hat{Y}^{HT}) = \sum_{s \in \Xi} \mathbf{p}(s) \hat{Y}_s^{HT} = Y . \tag{1.3}$$

The mean of the values of $\hat{Y}^{HT}$ weighted by the selection probability of sample s then corresponds to the true value of the total Y.

Consider t_k, an indicator variable where $t_k = 1$ if $k \in s$, and 0 otherwise. With this variable, we can rewrite the estimator $\hat{Y}^{HT}$ under the form

$$\hat{Y}^{HT} = \sum_{k=1}^{N} \frac{t_k}{\pi_k} y_k . \tag{1.4}$$

Moreover, we note that

$$E(t_k) = 1 \times P(k \in s) + 0 \times P(k \notin s) = P(k \in s) = \pi_k . \tag{1.5}$$

From (1.4) and (1.5), we can prove the unbiasedness of the Horvitz-Thompson estimator in the following way:

$$\begin{aligned} E(\hat{Y}^{HT}) &= E\left(\sum_{k=1}^{N} \frac{t_k}{\pi_k} y_k \right) = \sum_{k=1}^{N} \frac{E(t_k)}{\pi_k} y_k \\ &= \sum_{k=1}^{N} \frac{\pi_k}{\pi_k} y_k = \sum_{k=1}^{N} y_k = Y. \end{aligned} \tag{1.6}$$

The formula for the *variance* of the estimator $\hat{Y}^{HT}$, with respect to the sampling design, is given by

$$Var(\hat{Y}^{HT}) = \sum_{k=1}^{N} \sum_{k'=1}^{N} \frac{(\pi_{kk'} - \pi_k \pi_{k'})}{\pi_k \pi_{k'}} y_k y_{k'} \tag{1.7a}$$

or, in an equivalent manner, by

$$Var(\hat{Y}^{HT}) = -\frac{1}{2} \sum_{k=1}^{N} \sum_{k'=1}^{N} (\pi_{kk'} - \pi_k \pi_{k'}) \left(\frac{y_k}{\pi_k} - \frac{y_{k'}}{\pi_{k'}} \right)^2 \tag{1.7b}$$

---

from the population, we will respectively use $\Sigma_{i=1}^{N}$ and $\Sigma_{i=1}^{n}$. This notation has been used in several books on sampling theory such as, among others, Cochran (1977) and Morin (1993).

where $\pi_{kk'}$ represents the joint selection probability of units k and k'. For the details in the proofs of (1.7a) and (1.7b), we can consult Särndal, Swensson and Wretman (1992).

We can also write the estimator $\hat{Y}^{HT}$ given by (1.2) as a function of the *sampling weight* $d_k = 1/\pi_k$. We then have

$$\hat{Y}^{HT} = \sum_{k=1}^{n} d_k y_k . \tag{1.8}$$

In sampling theory, the sampling weight is the inverse of the selection probability π_k of unit k from sample s. The sampling weight of unit k corresponds to the expected number of units from population U represented by this unit. For example, if an individual has one chance out of four ($\pi_k = 1/4$) of being part of the sample, it will have a sampling weight of 4; we then say that this individual in the sample represents on average four individuals within the population. Let us note that the sampling weight d_k may possibly not be an integer.

It is possible to define in a general way an *estimation weight* w_k that we associate to unit k of sample s. This weight leads to the estimator

$$\hat{Y} = \sum_{k=1}^{n} w_k y_k . \tag{1.9}$$

The properties (bias and variance, for example) of this estimator depend upon the construction of the estimation weight w_k. In this book, we will focus on an estimation weight obtained by the generalisation of a method called weight share.

To learn more about sampling theory, the reader can consult books such as Cochran (1977), Grosbras (1986), Särndal, Swensson and Wretman (1992), Morin (1993), Ardilly (2006), and Lohr (1999).

1.2 CLUSTER SAMPLING

It often happens that sample surveys are performed in clusters. *Cluster sampling* is in fact a sampling design commonly used in practice. This technique of sampling is not suitable for the drawing of samples of units, but rather the selection of groups of units called *clusters* [or *primary sampling units* (PSU)]. The units in the clusters are called *secondary sampling units* (SSU). In cluster sampling, we survey for all the SSU belonging to the selected PSU. When we survey only for a subsample of the SSU, within the selected PSU, we are instead speaking of *two-stage sampling*.

For social studies, several surveys are built in such a way that we sample households in order to survey for the set of individuals from these households. The households thus form clusters of individuals. This is particularly the case for the Labour Force Survey conducted by Statistics Canada (Singh *et al.*, 1990). For economic surveys, the sampling of enterprises is often done with the goal of obtaining information on their components, for instance, the establishments or the local units. Enterprises are therefore composed of clusters of establishments, or local units, which we survey in order to provide economic statistics, in particular for national accounts.

With cluster sampling, the survey statistician can hope for reductions in collection costs. Indeed, surveying for entire households, for example, allows the interviewer to considerably reduce his number of trips compared to sampling for the same number of persons, but in different households. Cluster sampling also allows for the production of results at the cluster level itself, on top of the units. For example, we can calculate the average income of the households.

Cluster sampling is presented in most books that deal with sampling theory. We assume that the population U consists of N clusters where each cluster i contains M_i units. This is illustrated in Figure 1.1. We select a sample s containing n clusters in population U according to a certain sampling design. We assume that π_i represents

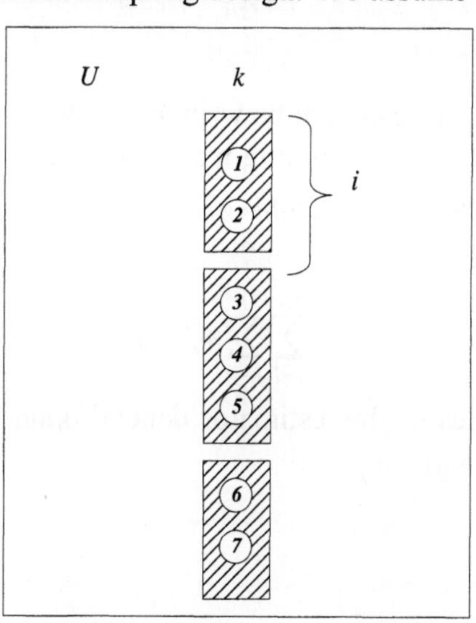

Figure 1.1: *Cluster sampling*

the selection probability of cluster *i*, where $\pi_i > 0$ for all clusters $i \in U$. As each cluster *i* of population *U* contains M_i units, we have in total $M = \sum_{i=1}^{N} M_i$ units in the population. We survey all units of clusters *i* for sample *s*. Each unit *k* of cluster *i* therefore has the same selection probability as the cluster, i.e., $\pi_{ik} = \pi_i$.

With cluster sampling, we are looking to estimate the total $Y = \sum_{i=1}^{N} \sum_{k=1}^{M_i} y_{ik}$ for a characteristic *y*. Considering the Horvitz-Thompson estimator (1.2), we can use the estimator $\hat{Y}^{CLUS,HT}$ given by

$$\hat{Y}^{CLUS,HT} = \sum_{i=1}^{n} \frac{Y_i}{\pi_i} \tag{1.10}$$

where $Y_i = \sum_{k=1}^{M_i} y_{ik}$. The superscript *CLUS* refers to the term *cluster sampling*. The variance of $\hat{Y}^{CLUS,HT}$ is given by

$$Var(\hat{Y}^{CLUS,HT}) = \sum_{i=1}^{N} \sum_{i'=1}^{N} \frac{(\pi_{ii'} - \pi_i \pi_{i'})}{\pi_i \pi_{i'}} Y_i Y_{i'} . \tag{1.11}$$

We can rewrite estimator (1.10) in the following manner:

$$\begin{aligned} \hat{Y}^{CLUS,HT} &= \sum_{i=1}^{n} \frac{1}{\pi_i} \sum_{k=1}^{M_i} y_{ik} \\ &= \sum_{i=1}^{n} \sum_{k=1}^{M_i} \frac{y_{ik}}{\pi_{ik}} = \sum_{i=1}^{n} \sum_{k=1}^{M_i} d_{ik} y_{ik} \end{aligned} \tag{1.12}$$

where $d_{ik} = 1/\pi_{ik}$.

Estimator (1.10) can then be written as a function of units *k* for clusters *i* of sample *s* with sampling weight d_{ik}. In a general way, we can construct an estimation weight w_{ik}^{CLUS} and define an estimator of the form

$$\hat{Y}^{CLUS} = \sum_{i=1}^{n} \sum_{k=1}^{M_i} w_{ik}^{CLUS} y_{ik} . \tag{1.13}$$

The properties of this estimator depend upon the construction of the estimation weight w_{ik}^{CLUS}.

1.3 INDIRECT SAMPLING

To select in a probabilistic way the necessary samples for social or economic surveys, it is useful to have available sampling frames, i.e., lists of units meant to represent the target populations. Unfortunately, it may happen that no available sampling frame corresponds directly to the desired target population. We can then choose a sampling frame that is indirectly related to this target population. We can thus speak of two populations U^A and U^B that are related to one another. We wish to produce an estimate for U^B but unfortunately, we only have a sampling frame for U^A. We can then imagine the selection of a sample from U^A and produce an estimate for U^B using the existing links between the two populations. This is what we can refer to as *indirect sampling.*

For example, consider the situation where the estimate is concerned with young children (units) belonging to families (clusters) but the only sampling frame we have is a list of parents' names. The target population is that of the children, but we must first select a sample of parents before we can select the sample of children. Note that the children of a particular family can be selected through the father or the mother.

This is illustrated by Figure 1.2. In this example, the families are represented by the rectangles and we note that the children can come from different unions.

Another example of an application of indirect sampling is the situation where we wish to conduct a survey of enterprises (clusters) when we only have an incomplete sampling frame of establishments of these enterprises. For each establishment selected from the sampling frame, we want to sample the set of establishments (units) belonging to the same enterprise. The establishments that are not represented in the frame must be represented by those that are part of this frame (Lavallée, 1998b).

This example can be represented by Figure 1.3. Here we see that establishments **a, b, c, d,** and **e** are part of the sampling frame whereas establishments **f** and **g** are not part of it.

A third example is one where we are looking to conduct a survey on people (units) who live in dwellings (clusters). We have for this case a sampling frame of dwellings, but which is unfortunately not up-to-date. This sampling frame does not contain, among others, renovations affecting the division of buildings into apartments.

An example of this type of renovation is illustrated in Figure 1.4a. We note that dwellings **a**, **b**, **c**, **d**, and **e** have been transformed to get dwellings **a**′, **b**′, **c**′, and **d**′. By selecting a sample of dwellings from the sampling frame, we then go to new dwellings using the correspondence between the old and new dwellings. This correspondence is illustrated in Figure 1.4b.

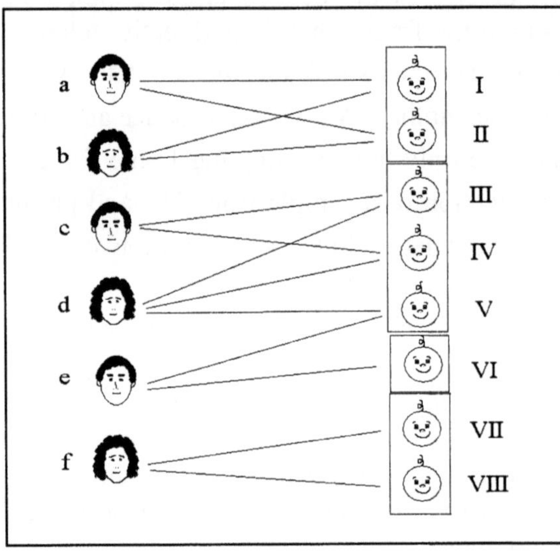

Figure 1.2: *Indirect sampling of children*

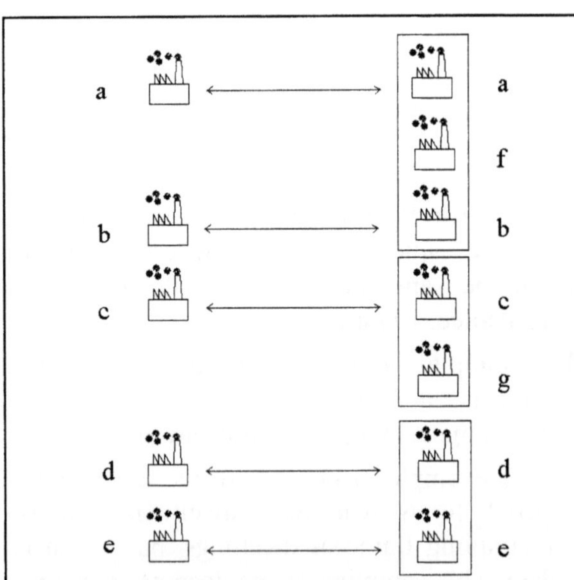

Figure 1.3: *Indirect sampling of*

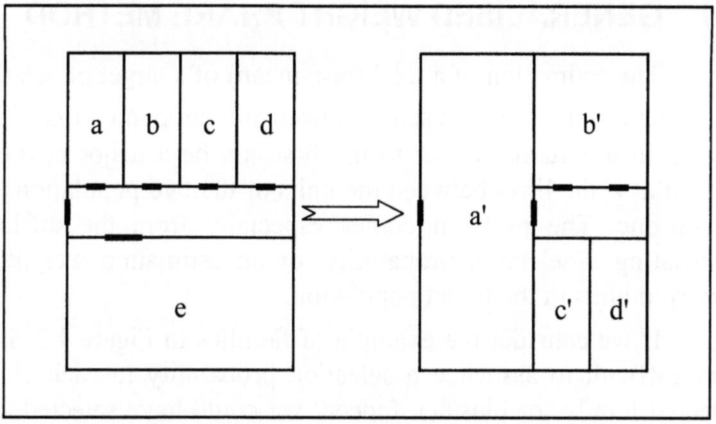

Figure 1.4a: *Indirect sampling of dwellings*

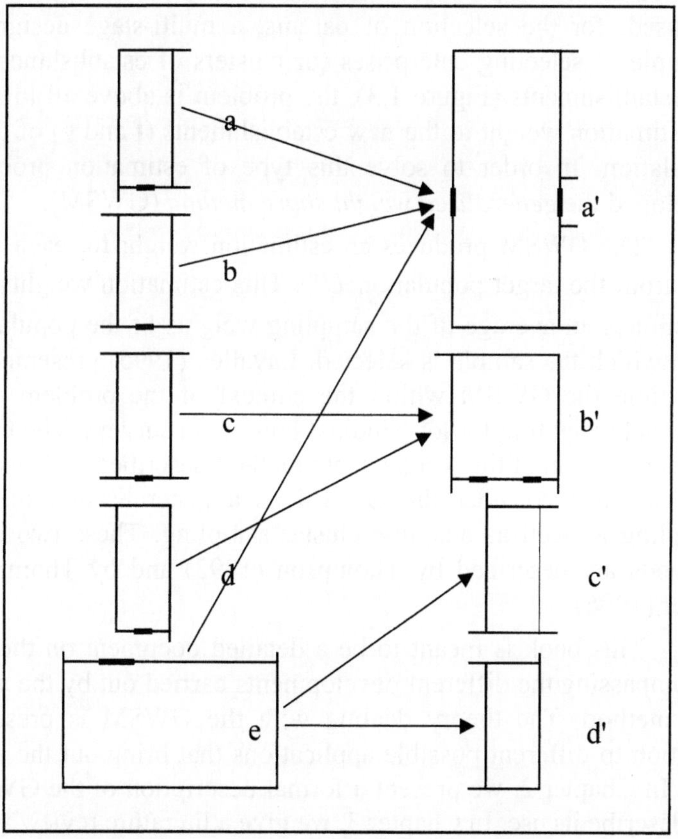

Figure 1.4b: *Indirect sampling of dwellings*

1.4 GENERALISED WEIGHT SHARE METHOD

The estimation of a total (or a mean) of a target population U^B of clusters using a sample selected from another population U^A that is related in a certain manner to the first can be a major challenge, in particular if the links between the units of the two populations are not one-to-one. The problem comes especially from the difficulty of associating a selection probability, or an estimation weight, to the surveyed units in the target population.

If we consider the example of families in Figure 1.2, it can be very difficult to associate a selection probability to each child of a selected family (or cluster). Indeed, we could have selected a family through one or more of the parents but, unfortunately, to know the selection probability of the family, and consequently of the children, we must know the selection probability of each parent, whether selected or not. In practice, this is not always the case, particularly if we used, for the selection of parents, a multi-stage design. In the example of selecting enterprises (or clusters of establishments) from the establishments (Figure 1.3), the problem is above all to associate an estimation weight to the new establishments (**f** and **g**) of the target population. In order to solve this type of estimation problem, we developed the *generalised weight share method* (GWSM).

The GWSM produces an estimation weight for each surveyed unit from the target population U^B. This estimation weight basically constitutes an average of the sampling weights of the population U^A from which the sample is selected. Lavallée (1995) presented for the first time the GWSM within the context of the problem of cross-sectional weighting for longitudinal household surveys. The GWSM is a generalisation of the weight share method described by Ernst (1989). We can also consider the GWSM as a generalisation of network sampling as well as adaptive cluster sampling. These two sampling methods are described by Thompson (1992) and by Thompson and Seber (1996).

This book is meant to be a detailed document on the GWSM encompassing the different developments carried out by the author on this method. The theory dealing with the GWSM is presented, in addition to different possible applications that bring out the appeal of this. In Chapter 2, we present a formal description of the GWSM and we describe its use. In Chapter 3, we give a literature review where we associate the GWSM with different sampling methods appearing in literature. We will see that the GWSM is a generalisation of methods such as the fair share method and adaptive cluster sampling. In Chapter 4, we present theoretical results on the GWSM, for instance the unbiasedness of the method and the variance of estimates resulting from it. In Chapter 5, we examine other possible generalisations of the GWSM. For example, we describe how to extend indirect sampling from one stage to two stages. In Chapter 6, we look at one of the main applications of the GWSM, being that related to longitudinal surveys. In Chapter 7, we describe how we can try to improve the precision of estimates coming from the GWSM by using calibration. In Chapter 8, we deal with the practical case where non-response occurs during data collection. We see that we can correct the weights coming from the GWSM by calculating a response probability associated with the responding units. In Chapter 9, we discuss the case where the links between populations U^A and U^B were established from a process of probabilistic linkage. We then see that it is possible to modify the GWSM in order to adapt it to the situation where the links between the two populations are not deterministic. Finally, we end the book with a conclusion that emphasises new applications of the indirect sampling.

CHAPTER 2

DESCRIPTION AND USE OF THE GWSM

As mentioned in the introduction, the GWSM was first described by Lavallée (1995). It produces an estimation weight for each unit surveyed in the target population U^B. This estimation weight basically constitutes an average of the sampling weights of the population U^A from which the sample is selected. We first present in this chapter a formal description of the GWSM. Second, we describe the use of the method.

2.1 DESCRIPTION

We select a sample s^A containing m^A units from the population U^A containing M^A units according to a certain sampling design. Suppose that π_j^A represents the selection probability of unit j. We assume that $\pi_j^A > 0$ for all $j \in U^A$. On the other hand, the target population U^B contains M^B units. This population is divided into N clusters,[1] where cluster i contains M_i^B units.

We assume there exists a *relationship* between units j of population U^A and units k of clusters i of the population U^B. This relationship is identified by an indicator variable $l_{j,ik}$, where $l_{j,ik} = 1$ if there exists a *link* between unit $j \in U^A$ and unit $ik \in U^B$, and 0 otherwise. Note that there might be some cases where no links exist

[1] We will use later the notation N^B (instead of N) to indicate the clusters of U^B as the population U^A itself will be divided into clusters. When we write N (and n) without superscripts, we will take for granted that it is a matter of the clusters of U^B

between a unit j of population U^A and units k of clusters i of population U^B, which comes back to saying that $L_j^A = \sum_{i=1}^{N} \sum_{k=1}^{M_i^B} l_{j,ik} = 0$. Moreover, there can be zero, one or more links for any unit k of a cluster i of population U^B, i.e., that it is possible to have $L_{ik}^B = \sum_{j=1}^{M^A} l_{j,ik} = 0$, $L_{ik}^B = 1$, or even $L_{ik}^B > 1$ for all units $ik \in U^B$. To use the GWSM, however, we must satisfy the following constraint.

Constraint 2.1 *Each cluster i of U^B must have at least one link with a unit j of U^A, i.e.,*

$$L_i^B = \sum_{k=1}^{M_i^B} \sum_{j=1}^{M^A} l_{j,ik} > 0.$$

We will see that this constraint is essential to ensure the unbiasedness of the GWSM.

For each unit *j* selected in s^A, we identify the units *ik* of U^B that have a non-zero link with *j*, i.e., $l_{j,ik} = 1$. If $L_j^A = 0$ for a unit *j* of s^A, there are simply no units of U^B identified by this unit *j*, which affects the efficiency of the sample s^A but does not introduce any bias. For each unit *ik* identified, we assume that we can set up the list of M_i^B units of cluster *i* containing this unit. Each cluster *i* then represents, within itself, a population U_i^B where $U^B = \bigcup_{i=1}^{N} U_i^B$. Let Ω^B be the set of *n* clusters identified by the units $j \in s^A$, i.e., $\Omega^B = \{i \in U^B \mid \exists j \in s^A \text{ and } L_{j,i} > 0\}$ where $L_{j,i} = \sum_{k=1}^{M_i^B} l_{j,ik}$.

We survey all the units *k* of clusters $i \in \Omega^B$ where we measure a certain variable of interest y_{ik} and the number of links $L_{ik}^B = \sum_{j=1}^{M^A} l_{j,ik}$ between unit *ik* of U^B and the population U^A. An important constraint to which the survey process (or measurement) is subjected is thus to consider **all** units within the same cluster. That is, if a unit is selected in the sample then every unit of the cluster containing the selected unit will be surveyed.

This constraint is one that often arises in surveys for two reasons: (i) cost reductions and (ii) the need for producing estimates on clusters. As an example, for social surveys, there is normally only a small marginal cost for interviewing all persons within the household. On the other hand, household estimates are often of special interest for those who are looking to measure poverty, for example.

Different cases of links are shown in Figure 2.1. Looking at it, we see that there is no unit j of U^A that does not have a link with U^B. Constraint 2.1 of the GWSM is satisfied here since all clusters i of U^B are linked to at least one unit j of U^A. This constraint is necessary to make the GWSM unbiased. We can indeed see that if a cluster does not have a link with U^A, it results in an underestimation of a total or mean of U^B since this cluster has no chance of being surveyed. As shown by unit 7 of U^B in Figure 2.1, it is possible for there to be no links for a given unit of a cluster i provided, however, that at least one unit of the cluster has a link with U^A, as claimed by Constraint 2.1.

For the target population U^B, we look to estimate the total $Y^B = \sum_{i=1}^{N} \sum_{k=1}^{M_i^B} y_{ik}$. By using the GWSM, we want to assign an *estimation weight* w_{ik} to each unit k of a surveyed cluster i. To estimate the total Y^B belonging to the target population U^B, we can then use the estimator

$$\hat{Y}^B = \sum_{i=1}^{n} \sum_{k=1}^{M_i^B} w_{ik} y_{ik} \qquad (2.1)$$

where n is the number of surveyed clusters and w_{ik}, the weight assigned to unit k of cluster i. With the GWSM, the estimation method is based on the sample s^A, together with the existing links between U^A and U^B to estimate the total Y^B. The links are in fact used as a bridge to go between the populations U^A and U^B.

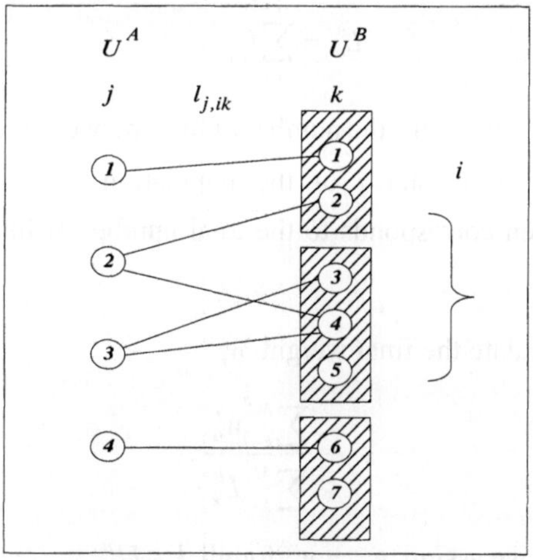

Figure 2.1: *Example of links*

The GWSM provides to each sampled unit a final weight calculated according to a weighted method within each cluster i entering into $\hat{Y}^B$. We first calculate an *initial weight* that corresponds to the inverse of the selection probability of units j of s^A that have a non-zero link with unit k of cluster i of $\hat{Y}^B$. An initial weight of zero is assigned to the units not having a link. The *final weight* is obtained by calculating the ratio of the sum of the initial weights for the cluster over the total number of links for that cluster. This final weight is finally assigned to all units within the cluster. Note that the fact of allocating the same estimation weight to all units has the considerable advantage of ensuring consistency of estimates for units and clusters.

Steps of the GWSM

Step 1: For each unit k of cluster i of Ω^B, we calculate the initial weight w'_{ik}, as follows:

$$w'_{ik} = \sum_{j=1}^{M^A} l_{j,ik} \frac{t_j}{\pi_j^A} \tag{2.2}$$

where $t_j = 1$ if $j \in s^A$, and 0 otherwise. Note that a unit ik having no link with any unit j of U^A automatically has an initial weight of zero.

Step 2: For each unit k of cluster i of Ω^B, we get the total number of links L_{ik}^B:

$$L_{ik}^B = \sum_{j=1}^{M^A} l_{j,ik} \,. \tag{2.3}$$

The quantity L_{ik}^B represents the number of links between the units of U^A and the unit k of cluster i of the population U^B. The quantity $L_i^B = \sum_{k=1}^{M_i^B} L_{ik}^B$ then corresponds to the total number of links present in cluster i.

Step 3: We calculate the final weight w_i:

$$w_i = \frac{\sum_{k=1}^{M_i^B} w'_{ik}}{\sum_{k=1}^{M_i^B} L_{ik}^B} \,. \tag{2.4}$$

Step 4: Finally, we assign $w_{ik} = w_i$ for all $k \in U_i^B$.

By following steps 1 to 4, we deduce the following result.

Result 2.1 *For the units ik of the target population U^B, we have*

$$w_{ik} = \sum_{j=1}^{M^A} \frac{t_j}{\pi_j^A} \frac{L_{j,i}}{L_i^B}.$$

Proof

$$\begin{aligned} w_{ik} &= \sum_{k=1}^{M_i^B} \frac{w'_{ik}}{L_i^B} = \sum_{k=1}^{M_i^B} \frac{1}{L_i^B} \sum_{j=1}^{M^A} \frac{t_j}{\pi_j^A} l_{j,ik} \\ &= \sum_{j=1}^{M^A} \frac{t_j}{\pi_j^A} \frac{1}{L_i^B} \sum_{k=1}^{M_i^B} l_{j,ik} = \sum_{j=1}^{M^A} \frac{t_j}{\pi_j^A} \frac{L_{j,i}}{L_i^B}. \end{aligned} \tag{2.5}$$

■

To estimate the total Y^B, we use equation (2.1). Because the estimation weights coming from the GWSM are the same for the set of M_i^B units of each cluster i, the estimator (2.1) can be written as a function of only clusters. Thus we have

$$\hat{Y}^B = \sum_{i=1}^{n} \sum_{k=1}^{M_i^B} w_{ik} y_{ik} = \sum_{i=1}^{n} w_i \sum_{k=1}^{M_i^B} y_{ik} = \sum_{i=1}^{n} w_i Y_i. \tag{2.6}$$

We will formally show in Chapter 4 that if Constraint 2.1 is respected, the estimator $\hat{Y}^B$ proves to be unbiased. Moreover, we can obtain a variance formula to calculate the precision of $\hat{Y}^B$.

***Example* 2.1**

As an example, take the case illustrated in Figure 2.1. We are here looking to estimate the total Y^B linked to the target population U^B.

Suppose that we select from U^A the unit j=1 and the unit j=2. Before applying the GWSM, we are going to re-index the units of U^B in accordance to the notation used in Figure 2.1. We thus obtain:

Units of U^B from Fig. 2.1	**1**	**2**	**3**	**4**	**5**	**6**	**7**
i	1	1	2	2	2	3	3
k	1	2	1	2	3	1	2

By selecting the unit j=1, we survey the units of cluster i=1. Likewise, by selecting the unit j=2, we survey the units of clusters i=1 and i=2. We therefore have $\Omega^B = \{1,2\}$. For each unit k of clusters i of Ω^B, we calculate the initial weight w'_{ik}, the number of links L_{ik}^B, and the final weight w_i, which gives us the table below.

i	k	w'_{ik}	L_{ik}^B	w_i
1	1	$\frac{1}{\pi_1^A}$	1	$\frac{1}{2}\left[\frac{1}{\pi_1^A}+\frac{1}{\pi_2^A}\right]$
1	2	$\frac{1}{\pi_2^A}$	1	$\frac{1}{2}\left[\frac{1}{\pi_1^A}+\frac{1}{\pi_2^A}\right]$
2	1	0 (because $t_3 = 0$)	1	$\frac{1}{3}\left[0+\frac{1}{\pi_2^A}+0\right]=\frac{1}{3\pi_2^A}$
2	2	$\frac{1}{\pi_2^A}+0=\frac{1}{\pi_2^A}$	2	$\frac{1}{3\pi_2^A}$
2	3	0 (because $l_{j,23} = 0$ for all j)	0	$\frac{1}{3\pi_2^A}$

The estimator $\hat{Y}^B$ given by (2.1) is finally written

$$\hat{Y}^B = \frac{1}{2}\left[\frac{1}{\pi_1^A}+\frac{1}{\pi_2^A}\right]y_{11} + \frac{1}{2}\left[\frac{1}{\pi_1^A}+\frac{1}{\pi_2^A}\right]y_{12} + \frac{y_{21}}{3\pi_2^A} + \frac{y_{22}}{3\pi_2^A} + \frac{y_{23}}{3\pi_2^A}.$$

2.2 USE

The GWSM derives its use in practical situations taking place in problems of sampling. It offers a simple solution to sampling problems and to complex weighting. Note that in simple problems concerning classical sampling theory, the GWSM in general gives the same results as classical theory. We will in particular verify this observation in section 4.2 for the case of cluster sampling.

In the world of surveys, it is often useful to have different relatively simple processes so as to minimise the possibility of errors that may arise during processing. With its simplicity, the GWSM offers an interesting solution that could be chosen, even though it could turn out that the GWSM is not the most precise (i.e., of minimal variance) compared to another more complex estimation method. This is particularly the case with the application of the GWSM in the Program to Improve Provincial Economic Statistics (PIPES) where the GWSM was used as the basis for calculating the variance of estimates. For more details, we can consult Girard and Simard (2000).

Here we present the four principal reasons to use the GWSM.

2.2.1 Indirect sampling for rare populations

In practice *rare populations* are often difficult to target for surveying purposes. Most of the time, we do not have any adequate sampling frames and we must therefore use a different but somewhat related sampling frame to reach the rare target population. An indirect sampling is thus performed. For example, to target people having some infectious disease in a large city, we can use lists of dwellings as sampling frames, which subsequently causes us to survey the families of the selected dwellings.

Fortunately for the statistician, it turns out that rare populations are often found in clusters. This is often the case, for example, with infectious diseases (Thompson, 1992). By surveying the complete clusters, we then see considerable reductions in costs since a large part of the costs are related to the identification of these rare populations. Therefore, in the end, we get data for the clusters of surveyed units through indirect sampling.

The problem for the statistician is finally to weight the survey data so that we can produce unbiased estimates for the characteristics of the rare target population. The GWSM provides a simple way of obtaining this weighting.

2.2.2 Weighting using only the selection probabilities of the selected units

The GWSM needs selection probabilities π_j^A only for the selected units *j* in the sample s^A. This is a major simplification compared to other weighting methods such as the one based on the exact calculation of selection probabilities of surveyed units. Take, for example, unit 2 of the population U^B from Figure 2.1. This unit is surveyed if we select the unit *j*=1, the unit *j*=2, or both, in sample s^A. Thus, we can in theory calculate the probability of surveying unit 2 that is approximately given by

$$\begin{aligned} P(\text{surveying unit 2}) &= P((j=1 \in s^A) \cup (j=2 \in s^A)) \\ &\approx 1-\left[1-P(j=1 \in s^A)\right] \left[1-P(j=2 \in s^A)\right] \\ &= 1-\left[1-\pi_1^A\right] \left[1-\pi_2^A\right]. \end{aligned} \tag{2.7}$$

Unfortunately, in practice, such a probability can be very difficult, if not impossible, to get. This is the case, for example, if sample s^A is selected from a multi-stage sampling design. With such a design, if we selected unit *j*=1 in the sample but not unit *j*=2, it is uncertain that we know the selection probability π_2^A (and vice versa), in particular if the two units *j*=1 and *j*=2 are not part of the same PSU. In this case, we cannot calculate probability (2.7) and we cannot then weight unit 2 from its probability of being surveyed. By only using the probabilities of selected units in sample s^A, the GWSM gives a simple solution that is applicable in all cases where we know the selection probabilities of units *j* in sample s^A.

2.2.3 Weighting of populations related by complex links

When we perform an indirect sampling, it often turns out that the links between the population U^A from which the sample is selected and the target population U^B are *complex*, that is to say, the links between U^A and U^B are of the type "many-to-many." For example, we can take the situation of the sampling of children of blended families illustrated in Figure 1.2. In this example, we select a sample of parents with the intention of surveying the children who belong to the families (clusters). If the two parents lived together with their children, we would be in a relatively simple case of conventional cluster sampling. However, in a case of blended families where the children living together are not necessarily brothers and sisters and where the parents do not necessarily live together, we find ourselves in a much more complex situation. To get an estimation weight for each child of the surveyed families, the GWSM then turns out to be very useful.

2.2.4 Weighting of unlinked units

Since here we are surveying the set of units from the clusters, it can happen that we must calculate an estimation weight for a unit of U^B that is surveyed but that is not linked to the population U^A from which the sample is selected. Such a situation is illustrated in Figure 2.1 by units 5 and 7.

A typical example of this type of situation comes from *longitudinal surveys* of individuals belonging to households. In this type of survey, we select a sample s^A of individuals from a population U^A. We then follow these individuals over time. During a second survey wave, following changes in the population (movements into and out of the population, modifications to the composition of households), we are faced with a new population U^B. The links between populations U^A and U^B here are associated to the individuals. The individuals of s^A can now belong to households that have individuals of U^A who have not been selected in s^A, or who are new (births or immigrants) to the population. Note that by definition, those who are new to the population do not have any links with U^A. Since we are surveying all individuals of the households having individuals from s^A, we thus get data for the new individuals. The problem is then to get an estimation weight for these units so that we can produce unbiased estimates, including the data from the new individuals. This problem, however, is not obvious to solve since the new individuals in the population were not selected at the time of s^A, but they are surveyed simply because they are part of households containing the individuals of s^A. Obtaining of an estimation weight for the new individuals in the population is, among others, discussed in Chapter 6. We will see that the GWSM allows for the finding of an elegant and unbiased solution for this problem.

CHAPTER 3

LITERATURE REVIEW

The GWSM turns out to be useful in the most diverse applications where we must obtain an estimate of a total for a population of clusters when meanwhile the sample comes from another population related to the first.

The first "official" application of the GWSM is that written by Lavallée (1995) where it was used to perform cross-sectional weighting for Statistics Canada's Survey of Labour and Income Dynamics (SLID). SLID is a longitudinal survey of individuals belonging to households (or clusters) where we must also produce cross-sectional estimates on top of longitudinal estimates. Owing to its importance, this example will be described in details in Chapter 6.

3.1 FIRST STEPS

Before its application in SLID, the foundations of the GWSM were already used to solve complex estimation problems. For example, the tax data program at Statistics Canada has used for many years a partner correction factor (PCF) to correct estimates in order to account for partners in a single business. These partners are in fact the tax filers who produce identical income tax reports for the same enterprise where they are in partnership as owners.

For the tax data program, a sample is selected from the file of tax filers (population U^A) to produce an estimate of gross income for the population of businesses (population U^B). Note that a tax filer can own many businesses and, in the case of partners, many tax filers can own the same business. If we consider the businesses as clusters, we are therefore, because of the "one-to-many" and "many-to-one" type of links, in a relatively complex case of estimation.

The PCF turns out to be a factor associated to the measured variables of the partners that reduces the value of these variables proportionally to the profits of the partners owning the business. For example, if a business has two owners and each partner earns 50% of the business profits, the measured variables of each partner, whether or not he is selected in the sample of tax filers, will be divided by two.

This case appears in the possible applications of the GWSM. We will expand this idea in section 4.5. Bankier (1983) showed that the PCF allowed for unbiased estimates to be produced.

3.2 FAIR SHARE METHOD

The *fair share method* can be considered as one of the precursors to the GWSM. This method was presented by Huang (1984), Judkins *et al.* (1984), Ernst, Hubble and Judkins (1984), and Ernst (1989), in the context of longitudinal surveys. These authors used the fair share method in order to solve theoretical and operation problems relative to the Survey of Income and Program Participation (SIPP) conducted in the United States for the Income Survey Development Program. SIPP is a longitudinal survey of persons and households. It is in fact similar to Statistics Canada's SLID that is described in details in Chapter 6 in the context of the application of the GWSM to longitudinal surveys.

Huang (1984) presented the fair share method with a method called *multiplicity approach* to solve the problem of cross-sectional weighting for longitudinal surveys of households. The problem is the following. In Wave 1, a sample of households containing persons is selected, and these persons are followed over time for SIPP. These persons are considered longitudinal. In a subsequent wave (say, Wave 2), the composition of households containing the longitudinal persons may have changed following departures, moves, marriages, births, etc. In each wave, **all** persons belonging to households containing longitudinal persons are surveyed. We again encounter the problem of surveying clusters of units (here persons) from the choice of one or more units of the cluster. The problem is then to associate to each household surveyed in Wave 2 an estimation weight so that we can produce unbiased estimates for the cross-sectional population of Wave 2. This problem is illustrated in Figure 3.1.

If we use the notation relative to the GWSM, we can formally describe the fair share method in the following manner. As we can see in Figure 3.1, waves 1 and 2 correspond, respectively, to the populations U^A and U^B. The population U^A is divided into N^A households (or clusters), where household ι contains M_ι^A persons (or units). Each household ι then represents, in itself, a population U_ι^A where $U^A = \bigcup_{\iota=1}^{N^A} U_\iota^A$.

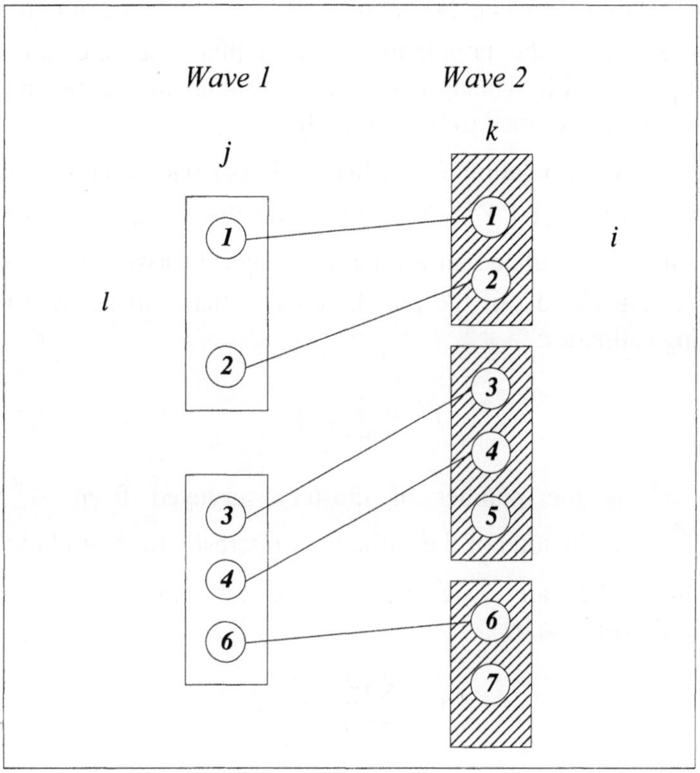

Figure 3.1: *Example of links in longitudinal surveys*

According to Huang (1984), a sample s^A of n^A households is selected among N^A from the population U^A following a certain sampling design. The sample contains in total m^A persons while the population U^A contains M^A persons. Let π_ι^A be the selection probability of household ι from U^A. Each person *j* of household ι has the same selection probability as its household. We assume that $\pi_\iota^A > 0$ for all households $\iota \in U^A$. The target population U^B corresponds to the same population as U^A, plus the persons who are added between waves 1 and 2. This population is divided into N^B households, where household *i* contains M_i^B persons.

In this context, the links between populations U^A and U^B are one-to-one for longitudinal persons and non-existent for persons who were added to the households (Figure 3.1). Huang (1984) noticed that there can be in practice households of U^B made up uniquely of persons added to the population U^A. In this case, the fair share method presented by Huang (1984) will produce an underestimate of the latter that is assumed to be negligible.

For the population U^B, Huang (1984) was interested in the estimation of the total $Y^B = \sum_{i=1}^{N}\sum_{k=1}^{M_i^B} y_{ik}$ for the variable of interest *y*. To do this, he calculated an estimation weight w_i associated with the surveyed household *i* of the population U^B that will be used in the following estimator:

$$\hat{Y}^B = \sum_{i=1}^{n^B} w_i Y_i, \tag{3.1}$$

where n^B is the number of clusters surveyed from U^B and $Y_i = \sum_{k=1}^{M_i^B} y_{ik}$ is the total of the variable of interest *y* for household *i*.

With the fair share method, the estimation weight w_i of household *i* is given by

$$w_i = \sum_{\iota=1}^{N^A} \frac{L_{\iota,i}}{L_i^B} \frac{t_\iota}{\pi_\iota^A}, \tag{3.2}$$

where $L_{\iota,i} = \sum_{j=1}^{M_\iota^A} \sum_{k=1}^{M_i^B} l_{\iota j,ik}$ and $L_i^B = \sum_{\iota=1}^{N^A} L_{\iota,i}$. The quantity $L_{\iota,i}$ is the number of links between household ι of U^A and household *i* of U^B. Because the links are one-to-one, $L_{\iota,i}$ corresponds to the number of persons from household *i* of Wave 2 coming from household ι of Wave 1. In a similar manner, the quantity L_i^B corresponds to the number of persons from Wave 1 belonging to household *i* of Wave 2.

We can show that the fair share method is only a particular case of the GWSM. Indeed, since we select entire households from Wave 1, we have $\pi_{\iota j}^A = \pi_\iota^A$ for all $j \in U_\iota^A$. Likewise, we have $t_{\iota j} = t_\iota$ for all $j \in U_\iota^A$. The initial weight w'_{ik}, given by (2.2) then takes the form:

$$w'_{ik} = \sum_{j=1}^{M^A} l_{j,ik} \frac{t_j}{\pi_j^A} = \sum_{\iota=1}^{N^A} \sum_{j=1}^{M_\iota^A} l_{\iota j,ik} \frac{t_{\iota j}}{\pi_{\iota j}^A} = \sum_{\iota=1}^{N^A} \frac{t_\iota}{\pi_\iota^A} \sum_{j=1}^{M_\iota^A} l_{\iota j,ik} = \sum_{\iota=1}^{N^A} \frac{t_\iota}{\pi_\iota^A} L_{\iota,ik}. \quad (3.3)$$

From (2.4), the final weight w_i is given by

$$w_i = \frac{\sum_{k=1}^{M_i^B} w'_{ik}}{\sum_{k=1}^{M_i^B} L_{ik}^B} = \frac{1}{L_i^B} \sum_{k=1}^{M_i^B} \sum_{\iota=1}^{N^A} \frac{t_\iota}{\pi_\iota^A} L_{\iota,ik} = \frac{1}{L_i^B} \sum_{\iota=1}^{N^A} \frac{t_\iota}{\pi_\iota^A} \sum_{k=1}^{M_i^B} L_{\iota,ik} = \frac{1}{L_i^B} \sum_{\iota=1}^{N^A} \frac{t_\iota}{\pi_\iota^A} L_{\iota,i}.$$

(3.4)

The estimation weight w_i obtained by the GWSM therefore corresponds exactly to that of the fair share method.

The method presented by Huang (1984) was named the fair share method because the estimation weight w_i given by (3.2) divides the value Y_i among the households of Wave 1 having contributed to household i of Wave 2, proportionally to the number of persons from the households involved. Indeed, by combining equations (3.1) and (3.2), we can rewrite the estimator $\hat{Y}^B$ in the following manner:

$$\hat{Y}^B = \sum_{i=1}^{n^B} \sum_{\iota=1}^{N^A} \frac{L_{\iota,i}}{L_i^B} \frac{t_\iota}{\pi_\iota^A} Y_i = \sum_{\iota=1}^{N^A} \frac{t_\iota}{\pi_\iota^A} \sum_{i=1}^{n^B} \frac{L_{\iota,i}}{L_i^B} Y_i. \quad (3.5)$$

We thus see that the value Y_i is divided according to the proportion of persons from household i of U^B coming from household ι of U^A with respect to the total number of persons from household i of U^B.

Huang (1984) presented, on top of the fair share method, the *multiplicity approach* that is, as a matter of fact, another approach to divide the value of Y_i. With this latest approach, the value of Y_i is divided according to the number of **households** η_i from Wave 1 having contributed (in terms of persons) to household i from Wave 2. The resulting estimator of Y^B is given by

$$\hat{Y}^{MULT,B} = \sum_{\iota=1}^{N^A} \frac{t_\iota}{\pi_\iota^A} \sum_{i=1}^{n^B} \frac{\delta_{\iota,i}}{\eta_i} Y_i, \quad (3.6)$$

where $\delta_{\iota,i} = 1$ if $L_{\iota,i} > 0$, and 0 otherwise. The indicator variable $\delta_{\iota,i}$ denotes whether or not household ι of U^A contributed to household i of U^B. Note that $\eta_i = \sum_{\iota=1}^{N^A} \delta_{\iota,i}$.

Huang (1984) proved that the fair share method and the multiplicity approach are both unbiased for the estimate of the total

Y^B. Note that the unbiasedness of the fair share method also directly follows from the fact that this is an application of the GWSM, whose unbiasedness will be shown in Chapter 4.

From an operational point of view, Huang (1984) mentioned that the fair share method is more appealing than the multiplicity approach because the quantities $L_{\iota,i}$ and L_i^B going into (3.2) are easier to obtain than the quantities $\delta_{\iota,i}$ and η_i going into (3.6). Indeed, after many waves, it can be difficult to know how many different households from Wave 1 contributed to a given household i of the current wave. However, it is relatively easy to know how many persons from Wave 1 contributed to household i of the current wave because the persons (and not the households) are followed over time. Note that the two approaches are identical if we assume that the households ι of U^A are of size 1.

In addition to the operational aspect, the fair share method also seems to have an advantage in the precision of the estimate of Y^B. Under certain hypotheses, Huang (1984) gave a heuristic proof that the fair share method is of minimal variance compared to any other method for dividing the value of Y_i. This speaks in favour of the GWSM as the fair share method is only a particular case of the GWSM.

3.3 CONTRIBUTION OF ERNST (1989)

Ernst (1989) presented a form of generalisation of the fair share method of Huang (1984). This method, based on the calculation of an average of weights for the cross-sectional weighting of individuals belonging to households, can be called the *weight share method* (WSM). The WSM differs notably from the fair share method in that the calculation of the estimation weights is centred on the individuals rather than the households. Note, however, that in the majority of applications, the two methods give identical results. The article by Ernst (1989) as a matter of fact acted as a basis for the GWSM.

The formal description of the WSM is written in the same context as that of Huang's fair share method (1984). Waves 1 and 2 correspond to the populations U^A and U^B from Figure 3.1. The population U^A is divided into N^A households (or clusters), where household ι contains M_ι^A persons (or units). We select a sample s^A of n^A households among the N^A from population U^A according to a certain sampling design. This sample contains in total m^A persons whereas population U^A contains M^A persons. Let $\pi_\imath^A$ be the selection probability of household $\imath$ from U^A where we assume that $\pi_\imath^A > 0$ for all households $\imath \in U^A$. Each person *j* of household $\imath$ has the same selection probability as the household and thus $\pi_{\imath j}^A = \pi_\imath^A$ for all units *j* of household $\imath$. The population U^B corresponds to the same population as U^A, plus the persons who are added between waves 1 and 2. In this way, the links between populations U^A and U^B are one-to-one (Figure 3.1). The population U^B is divided into N^B households, where household *i* contains M_i^B persons. We assume that sample s^A led to a survey of n^B households within the target population U^B.

By applying the WSM, we want to assign an estimation weight w_{ik}^{MPP} to each unit *k* of a surveyed cluster *i*. To estimate the total Y^B of the target population U^B, we can then use

$$\hat{Y}^{WSM,B} = \sum_{i=1}^{N^B} \sum_{k=1}^{M_i^B} w_{ik}^{WSM} y_{ik}, \tag{3.7}$$

where w_{ik}^{WSM} is the weight assigned to unit *k* of cluster *i*. This weight is 0 for the $N^B - n^B$ clusters *i* of U^B that are not surveyed.

Ernst (1989) mentioned that in the classical approach of a cross-sectional survey, the estimation weight w_{ik}^{WSM} for the surveyed units corresponds to the inverse of the selection probability of unit *ik*. We then produce unbiased estimates for the total Y^B. In the context of longitudinal surveys, the selection probability of surveyed units (or persons) can be difficult, indeed impossible, to obtain. This problem arises for persons who are surveyed simply because they live in households having persons from sample s^A (see Chapter 6).

Judkins *et al.* (1984), Ernst, Hubble and Judkins (1984), and Ernst (1989) noted that to produce unbiased estimates of the total Y^B, it is not necessary to know all the selection probabilities of the units *ik* going into (3.7). A necessary condition is simply to have

$$E(w_{ik}^{WSM}) = 1 \tag{3.8}$$

for all M^B units of U^B. Indeed, if $E(w_{ik}^{WSM}) = 1$, we have

$$E(\hat{Y}^{WSM,B}) = \sum_{i=1}^{N^B}\sum_{k=1}^{M_i^B} E(w_{ik}^{WSM})y_{ik} = \sum_{i=1}^{N^B}\sum_{k=1}^{M_i^B} y_{ik} = Y^B . \qquad (3.9)$$

The WSM allows us to get estimation weights that satisfy equation (3.8). We now show the steps given by Ernst (1989) for obtaining these weights.

Steps of the WSM

Step 1: For each unit k of cluster i, calculate the initial weight w'^{WSM}_{ik}, that is:

$$w'^{WSM}_{ik} = \begin{cases} 1/\pi^A_{ik} & \text{for units selected in } s^A \\ 0 & \text{otherwise} \end{cases} \qquad (3.10)$$

Step 2: Define constants α_{ik}, i=1,...,N^B and k=1,..., M_i^B. These constants are independent of the initial weights w'^{WSM}_{ik} and we have:

$$\sum_{k=1}^{M_i^B} \alpha_{ik} = 1 . \qquad (3.11)$$

Step 3: Calculate the final weight w_i^{WSM}:

$$w_i^{WSM} = \sum_{k=1}^{M_i^B} \alpha_{ik} w'^{WSM}_{ik} . \qquad (3.12)$$

Step 4: Set $w_{ik}^{WSM} = w_i^{WSM}$ for all units k of clusters i.

The constants α_{ik} form a sort of generalisation of the fair share method of Huang (1984). By assigning certain α_{ik} to zero, we can exclude some people from the calculation of the final weights given by (3.12). For example, we can decide to exclude from the weighting people less than 16 years of age by assigning them $\alpha_{ik} = 0$. Ernst (1989) gave different possible choices for the α_{ik} for the cross-sectional weighting of longitudinal surveys for individuals belonging to households. Ernst (1989), among other things, noticed that the most common choice for the constants α_{ik} is that where these constants correspond to the inverse of the number of persons M_i^{AB} from household i who belong to the two populations U^A and U^B.

We can see that the GWSM can constitute a generalisation of the WSM. The GWSM in fact elaborates beyond the context of longitudinal surveys since it allows the use of links that are not necessarily one-to-one between the populations U^A and U^B. In the context of the GWSM, since the links between populations U^A and U^B here are one-to-one for the longitudinal persons, we have $l_{j,ik} = 1$ if person *j* of U^A corresponds to person *k* of cluster *i* from U^B, and 0 otherwise. The indices *j* and *ik* are thus interchangeable for persons belonging to the two populations. The initial weight (2.2) is then

$$w'_{ik} = \sum_{j=1}^{M^A} l_{j,ik} \frac{t_j}{\pi_j^A} = \frac{t_{ik}}{\pi_{ik}^A} = w_{ik}'^{WSM} . \tag{3.13}$$

Likewise,

$$\sum_{k=1}^{M_i^B} L_{ik}^B = L_i^B = M_i^{AB} . \tag{3.14}$$

If we concentrate on the choice $\alpha_{ik} = 1/M_i^{AB}$, by replacing the quantities (3.13) and (3.14) in expression (2.4) for the final weight of the GWSM, we directly obtain

$$w_i = \frac{\sum_{k=1}^{M_i^B} w_{ik}'^{WSM}}{M_i^{AB}} = \sum_{k=1}^{M_i^B} \alpha_{ik} w_{ik}'^{WSM} = w_i^{WSM} . \tag{3.15}$$

Kalton and Brick (1995) as well as Lavallée and Deville (2002) studied the determination of optimal values for the constants α_{ik}. Because the problem turns out to be relatively complex to solve, Kalton and Brick particularly concentrated on the case where two households of U^A form a new household (or cluster) *i* of U^B. They drew the following conclusion: "in the two-household case, the equal household weighting scheme minimizes the variance of the household weights around the inverse selection probability weight when the initial sample is an epsem one". Minimising the variance of the household weights corresponds here to minimising the variance of the estimate of Y^B. What Kalton and Brick called the *equal household weighting scheme* is in fact the multiplicity approach described by Huang (1984) and presented in Section 3.2. Recall that with this approach, the weighting is calculated by dividing according to the number of **households** of U^A having contributed (in terms of persons) to household *i* of U^B. They add that "in the case of an approximately epsem[1] sample, the equal household weighting scheme should be close to the optimal, at least for the case where the members of the household at time t come from one or two households at the initial wave." These conclusions do not directly support the WSM (and consequently the GWSM) since this latter conclusion is quite similar to an *equal person weighting scheme*. Indeed, with the WSM, the weighting is calculated by dividing according to the number of **persons** M_i^{AB} from U^A having contributed to household i of U^B. Note however that if s^A is a sample of persons, considering the fact that the persons represent households of size 1, the equal weighting of households and the equal weighting of persons are equivalent. More recently, Deville and Lavallée (2006) obtained the necessary and sufficient conditions to obtain optimal weights for the GWSM. Their results are presented in details in Section 4.6.3.

Like Huang (1984), Kalton and Brick (1995) recognised that the WSM is more interesting in practice than the equal household weighting scheme (or the multiplicity approach). With the multiplicity approach, we need to know the number of households of U^A that provided the persons of a household i from U^B, which is sometimes difficult to establish. Thus, it can be difficult to know if two people from a household i of U^B live in the same household $\imath$ of U^A. So, although it might not be completely optimal, the GWSM offers an interesting solution, especially from the practical point of view, for the case of longitudinal surveys.

3.4 NETWORK SAMPLING

Network sampling is a survey method often used in social surveys. It proves to be particularly useful, for example, in defining populations that are rare or difficult to identify. In this type of sampling, the notion of *network* often corresponds to a range or set of contacts. We select units called *enumeration units* and we ask them to mention persons that they know corresponding to the desired criteria. We can illustrate the use of network sampling from an application by Sanders and Kalsbeek (1990). They used network sampling for a survey of pregnant women taken from a list of telephone numbers. The procedure consisted of contacting by telephone a certain number of persons selected by random digit dialling. They were then asked to mention if they knew any pregnant women among their family or

[1] equal-probability-of-selection method

friends. Sanders and Kalsbeek (1990) tested different options for the set of contacts (children, brothers and sisters; brothers, sisters, uncles and aunts; brothers, sisters and cousins, etc.) and it turned out that none of the options was really better than the others.

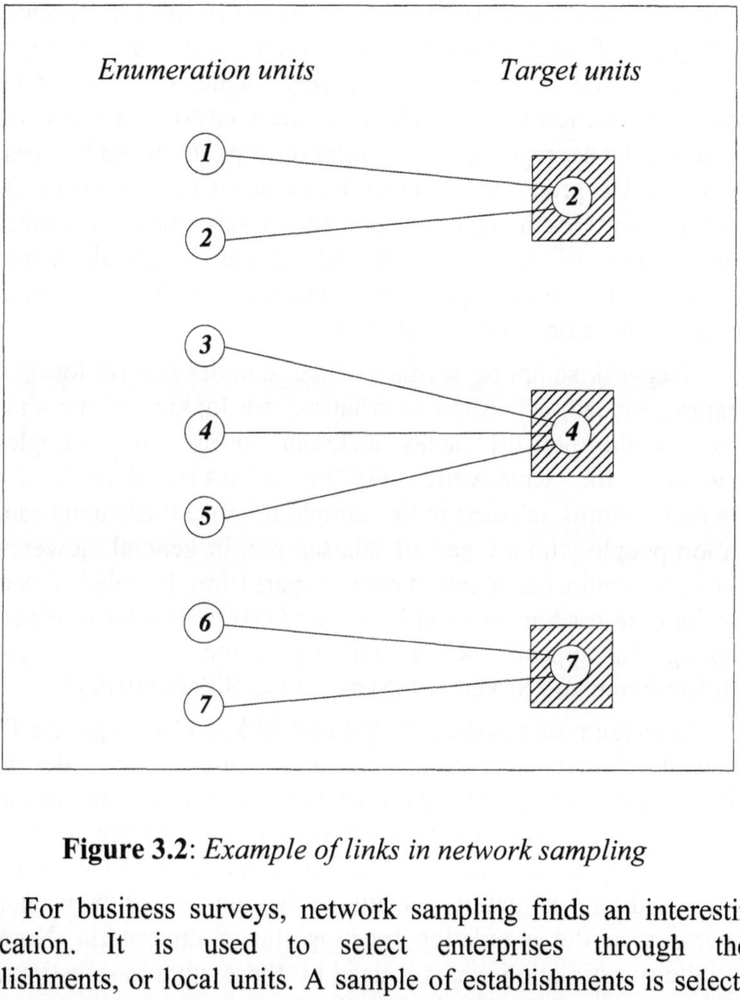

Figure 3.2: *Example of links in network sampling*

For business surveys, network sampling finds an interesting application. It is used to select enterprises through their establishments, or local units. A sample of establishments is selected and subsequently, enterprises having the selected establishments are surveyed. This example is illustrated by Figure 3.2. This method is used in particular at Statistics Canada within the Project to Improve Provincial Economic Statistics (PIPES). For more details, we can refer to Girard and Simard (2000).

One of the problems with the application of network sampling for social surveys is the difficulty of defining the network itself, i.e., the desired set of contacts. In the case of family relationships, this set is clearly defined, though its definition remains relatively arbitrary. Indeed, we can decide, for example, to include or exclude the grandparents from the family set. As mentioned by Sirken and Levy (1974), the definition of the network influences the selection probability of the target units, and thus affects the precision of the resulting estimates. In certain cases, the network itself is misspecified. For example, if we ask the selected people to "mention the people that they know," the set becomes relatively vague. Granovetter (1976) proposed a solution to this problem in the context of a study looking to measure the average number of relationships (or acquaintances) in a population. He proposed to select a sample of persons (enumeration units) and to ask each person if he or she knows, one after another, the other members of the list of selected persons. Although interesting, this solution becomes impossible to put into practice as soon as the sample reaches about a hundred people.

Network sampling seems to emerge under several forms in the literature. Indeed, according to what we are looking to measure, the notion of the network takes different forms. For example, the "networks" for Granovetter (1976) is restricted only to the enumeration units selected in the sample because these units can only mention people who are part of this sample. In general, however, the enumeration units can mention people apart from the selected persons, as is the case for Sanders and Kalsbeek (1990). The form of network sampling that appears most commonly in literature is that coming from Birnbaum and Sirken (1965) as well as Sirken (1970).

Birnbaum and Sirken (1965) and Sirken (1970) gave a formal statistical framework about network sampling by developing *multiplicity estimation.* This form of estimation takes into account the number of times a targeted person was mentioned by the enumeration units. Sirken (1970), Sirken (1972), Sirken and Levy (1974), and Levy (1977) used this estimation to evaluate the number (or the proportion) of persons in the population meeting the given criteria. Note that multiplicity estimation was not used by these authors to estimate the totals of quantitative variables. Multiplicity estimation certainly contributed to inspire the multiplicity approach described by Huang (1984).

Following the notation used for the GWSM, we can describe in a formal manner multiplicity estimation, and, in the process, network sampling. As seen in Figure 3.2, the enumeration units form the population U^A whereas the target units — those which have the desired characteristics — form the population U^B. Note that Sirken

(1970) assumed that the population U^A is a population of households, i.e., that each enumeration unit *j* of U^A corresponds to a household. Although the GWSM generally considers the units of U^A as being simple units (people, local units, etc.), this does not change anything in the theory presented.

According to Sirken (1970), a simple random sample s^A of m^A enumeration units is selected from the population U^A containing M^A units. Each enumeration unit j therefore has the same selection probability $\pi_j^A = m^A / M^A$. The population U^B corresponds to that of the target units, i.e., the units that have the desired characteristics. The population U^B has M^B target units.

As for the GWSM, Sirken (1970) used an indicator variable *l* to denote the link between enumeration units of U^A and target units of U^B. Therefore, we have $l_{j,k}=1$ if enumeration unit $j \in U^A$ identifies target unit $k \in U^B$, and 0 otherwise. We see here that the links are often many-to-one between U^A and U^B.

Sirken (1970) was interested in the estimation of the population count M^B of the target population U^B. For example, we can think of the estimation of the total number of pregnant women in a given region, as in the application of Sanders and Kalsbeek (1990). To do this, he calculated the following *multiplicity weight* ω_j, associated to each unit *j* selected in s^A:

$$\omega_j = \sum_{k=1}^{M^B} \frac{l_{j,k}}{L_k^B} \tag{3.16}$$

where $L_k^B = \sum_{j=1}^{M^A} l_{j,k}$. The multiplicity weight ω_j is so named because it keeps count of the number of times L_k^B that target unit k can be mentioned by the different enumeration units of U^A.

The multiplicity estimator $\hat{M}^{NET,B}$ of M^B is finally given by:

$$\hat{M}^{NET,B} = \sum_{j=1}^{m^A} \frac{\omega_j}{\pi_j^A} \tag{3.17}$$

where the superscript "NET" refers to network sampling.

It is relatively simple to show that multiplicity estimation, and thus network sampling, is a particular case of the GWSM. Although the target population U^B does not contain clusters as such, we can assume that the target units k of U^B in fact belong to clusters of size 1. In the context of the GWSM, we can then ignore the index i. Since we are interested here in the estimation of a population count, the variable of interest y here simply takes the value 1. The initial weight w'_{ik} given by (2.2) takes the form:

$$w'_k = \sum_{j=1}^{M^A} l_{j,k} \frac{t_j}{\pi_j^A}. \tag{3.18}$$

From (2.4), the final weight w_k is given by

$$w_k = \frac{w'_k}{L_k^B} = \frac{1}{L_k^B} \sum_{j=1}^{M^A} l_{j,k} \frac{t_j}{\pi_j^A}. \tag{3.19}$$

To estimate M^B, estimator (2.1) can then be written as

$$\hat{Y}^B = \sum_{k=1}^{m^B} w_k = \sum_{k=1}^{m^B} \frac{1}{L_k^B} \sum_{j=1}^{M^A} l_{j,k} \frac{t_j}{\pi_j^A} = \sum_{j=1}^{M^A} \frac{1}{\pi_j^A} \sum_{k=1}^{m^B} \frac{l_{j,k} t_j}{L_k^B}. \tag{3.20}$$

Following the survey process, the m^B target units are surveyed if and only if there is a link between unit j of U^A and k of U^B, and $t_j = 1$ (unit j of U^A is selected in s^A). In other words, the target unit k is surveyed if and only if we have $l_{j,k} t_j \neq 0$. The m^B surveyed units have therefore $l_{j,k} t_j / L_k^B \neq 0$, and the $M^B - m^B$ unsurveyed units have $l_{j,k} t_j / L_k^B = 0$.[2] Thus,

$$\hat{Y}^B = \sum_{j=1}^{M^A} \frac{1}{\pi_j^A} \sum_{k=1}^{m^B} \frac{l_{j,k} t_j}{L_k} = \sum_{j=1}^{M^A} \frac{t_j}{\pi_j^A} \sum_{k=1}^{M^B} \frac{l_{j,k}}{L_k^B} = \sum_{j=1}^{M^A} \frac{t_j}{\pi_j^A} \omega_j = \sum_{j=1}^{m^A} \frac{\omega_j}{\pi_j^A} = \hat{M}^{NET,B}. \tag{3.21}$$

Sirken (1972) extended multiplicity estimation in the case where the sample s^A is no longer a simple random sample but rather a stratified sample. Owing to the generality of the GWSM, it is simple to show again that multiplicity estimation is just an application of the GWSM.

Sirken (1970) and Sirken (1972) showed that multiplicity estimation is unbiased. With respect to the precision of estimates,

---

[2] Note that a similar argument will be used in Chapter 4 in the proof of Theorem 4.1.

network sampling seems to have an advantage compared to conventional sampling where each enumeration unit only reports for itself. Indeed, under certain conditions, Sirken (1970) showed that multiplicity estimation, and thus network sampling, can give inferior variances to those coming from estimators used in conventional surveys. Again, this speaks in favour of the GWSM since we showed that multiplicity estimation is nothing more than a particular case of the GWSM.

3.5 ADAPTIVE CLUSTER SAMPLING

Thompson (1992) and Thompson (2002) discussed sampling methods to use for populations that are difficult to reach because there is no sampling frame or because these populations are migratory or elusive. We can think, for example, of the problem of counting populations of fish in a lake, the assessment of the number of trees in a forest, or even the estimation of the number of people belonging to certain target groups (a particular ethnic origin or a socioprofessional category, for example) in a city. To solve this type of problem, Thompson (1990) proposed *adaptive cluster sampling.*

Adaptive cluster sampling is similar to network sampling and is particularly used to produce estimates for populations that are difficult to reach. Suppose, for example, that we are looking to estimate the number of people in a city, having income greater than $200,000. First of all, note that it is strongly possible that people with similar income live in the same neighbourhoods. To estimate this population count, we first select a small number of units (for example, houses) and we measure the income (left table of Figure 3.3). If a unit has income greater than $200,000, we then go to see the contiguous neighbours of this unit and also measure their income. For the new units where we found income greater than $200,000, we go to see their neighbors, and so on until we find no more neighbours with $200,000 in income (right table of Figure 3.3). We can thus obtain a considerable sample with, *a priori*, very little information about the units of the target population. Note that the sample is modified (or adapted) as the interviews progress.

With adaptive cluster sampling, the final clusters containing the target units are not distinct. This is due to the *edge units* that are units adjacent to the clusters of target units but are not part of them. Let us come back to the example of the measurement of income greater than $200,000, and assume that a house **a** neighbouring a targeted house **b** does not have income greater than $200,000. If house

a is selected, the process of adaptive cluster sampling will stop there because unit **a** does not belong to the target population. On the other hand, by selecting unit **b**, the survey process will continue and unit **a** will be surveyed because it is adjacent to **b**. The edge unit **a** can thus be surveyed in two clusters. Nevertheless, note that it will not contribute to the estimates because it is not part of the target population. Thompson (1990) bypasses this problem by defining *networks* that are in fact the final clusters, excluding edge units. These latter cases form the networks of size 1.

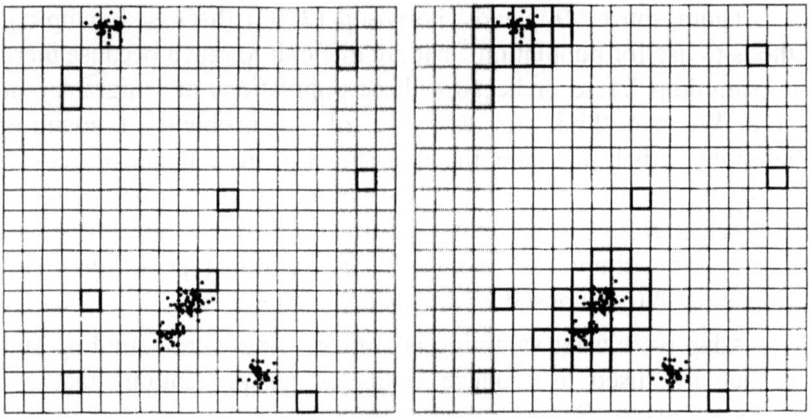

Figure 3.3: *Example of adaptive cluster sampling*

The networks are mutually exclusive and exhaustive. Whichever units are selected in the starting sample, we will have the same composition of networks at the end of the survey process. This comes from the fact that the established procedure to identify the "neighbours" of the selected units is independent from the selection process. Adaptive cluster sampling is therefore only a form of cluster sampling where the clusters here are networks selected from their component units. This type of sampling is often employed in practice. For example, in social surveys, it happens that people are selected from a list and subsequently all the people from their households are surveyed (Lavallée, 1995). At the business survey level, we often decide to select a sample of establishments (or local units) to then go up to the enterprise level to finally survey all establishments of this enterprise. Such a procedure is described, among others, in Lavallée (1998a).

Adaptive cluster sampling was described in details by Thompson (1990), Thompson (1991a), Thompson (1991b), Thompson

(1992), Thompson and Seber (1996), and Thompson (2002). We now present in a formal manner adaptive cluster sampling following the notation used for the GWSM. In this type of sampling, the populations U^A and U^B in fact correspond to the same population; the difference being that population U^B is formed by networks (or clusters, if we ignore the edge units). This is illustrated in Figure 3.4. We note that the subscripts *j* and *k* refer to the same units.

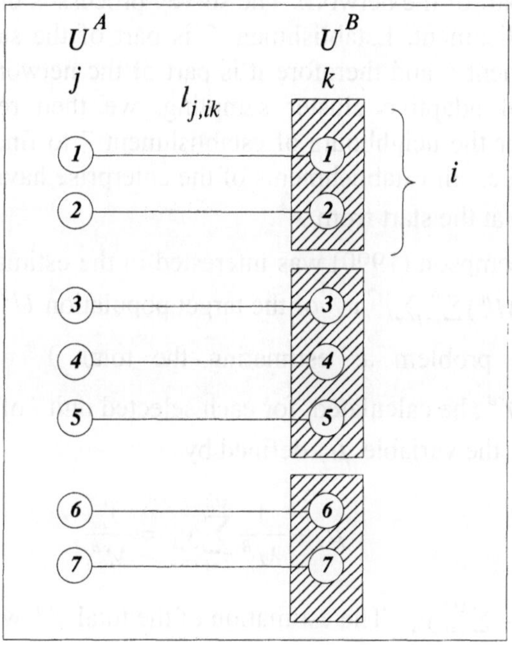

Figure 3.4: *Example of links in adaptive cluster sampling*

According to Thompson (1990), a sample s^A is selected containing m^A units in the population U^A containing M^A units using a certain sampling design. Assume that π_j^A represents the selection probability of unit *j* and that $\pi_j^A > 0$ for all $j \in U^A$. The target population U^B contains M^B units, where $M^A = M^B$. This population is divided into *N* networks, where network *i* contains M_i^B units.

Once the sample s^A is selected, the units j of s^A are surveyed. As shown in Figure 3.4, this corresponds to surveying the units k of U^B associated to the units j of s^A. The process of adaptive cluster sampling then requires going to survey the "neighbours" of the selected units. Let us again take the example of selecting enterprises through its establishments and assume that establishment 6 from Figure 3.4 was selected. Here, the neighbours are establishments 5 and 7. By surveying establishment 5, we realise that it is not part of the same enterprise as establishment 6, and it is therefore not considered as being part of the network. The survey process is then terminated for this establishment. Establishment 7 is part of the same enterprise as establishment 6 and therefore it is part of the network. Following the process of adaptive cluster sampling, we then restart the survey process for the neighbours of establishment 7 to finally complete the network, i.e., all establishments of the enterprise having establishment 6 selected at the start from s^A.

Thompson (1990) was interested in the estimation of the mean $\bar{Y}^B = (1/M^B)\sum_{i=1}^{N}\sum_{k=1}^{M_i^B} y_{ik}$ of the target population U^B, which is in fact the same problem as estimating the total $Y^B = \sum_{i=1}^{N}\sum_{k=1}^{M_i^B} y_{ik}$. To estimate Y^B, he calculated, for each selected unit j of s^A and linked to network i, the variable μ_j defined by

$$\mu_j = \frac{1}{M_i^B}\sum_{k=1}^{M_i^B} y_{ik} = \frac{Y_i}{M_i^B}, \tag{3.22}$$

where $Y_i = \sum_{k=1}^{M_i^B} y_{ik}$. The estimation of the total Y^B was then given by

$$\hat{Y}^{ADAP,B} = \sum_{j=1}^{m^A} \frac{\mu_j}{\pi_j^A}, \tag{3.23}$$

where the superscript "ADAP" refers to adaptive cluster sampling.

We can show that adaptive cluster sampling is just a particular case of the GWSM. First of all, recall that each unit j of U^A corresponds to a unit k from the network (or cluster) i of U^B. Consequently, $l_{j,ik} = 1$ for $j=ik$, and 0 otherwise. We can thus interchange the indices j and ik. Furthermore, $L_{ik}^B = \sum_{j=1}^{M^A} l_{j,ik} = 1$.

For each unit k of the cluster i going into $\hat{Y}^B$, the initial weight (2.2) here is given by:

$$w'_{ik} = \frac{t_{ik}}{\pi^A_{ik}}. \tag{3.24}$$

Here, the final weight w_i given by (2.4) takes the form:

$$w_i = \frac{\sum_{k=1}^{M_i^B} w'_{ik}}{\sum_{k=1}^{M_i^B} L^B_{ik}} = \frac{1}{M_i^B} \sum_{k=1}^{M_i^B} \frac{t_{ik}}{\pi^A_{ik}}. \tag{3.25}$$

From $\hat{Y}^B = \sum_{i=1}^n w_i \sum_{k=1}^{M_i^B} y_{ik} = \sum_{i=1}^n w_i Y_i$, we substitute the definition of w_i in $\hat{Y}^B$ to obtain

$$\hat{Y}^B = \sum_{i=1}^n \frac{Y_i}{M_i^B} \left(\sum_{k=1}^{M_i^B} \frac{t_{ik}}{\pi^A_{ik}} \right). \tag{3.26}$$

Following the survey process of adaptive cluster sampling, the M_i^B units of network i are surveyed if $t_{ik} = t_j = 1$ (unit j of U^A linked to unit *ik* of U^B is selected in s^A) for at least one $k \in U_i^B$. The n surveyed networks have thus $\sum_{k=1}^{M_i^B} t_{ik} / \pi^A_{ik} \neq 0$, and the $N-n$ unsurveyed networks have $\sum_{k=1}^{M_i^B} t_{ik} / \pi^A_{ik} = 0$. So,

$$\begin{aligned} \hat{Y}^B &= \sum_{i=1}^n \frac{Y_i}{M_i^B} \left(\sum_{k=1}^{M_i^B} \frac{t_{ik}}{\pi^A_{ik}} \right) \\ &= \sum_{i=1}^N \frac{Y_i}{M_i^B} \left(\sum_{k=1}^{M_i^B} \frac{t_{ik}}{\pi^A_{ik}} \right) \\ &= \sum_{i=1}^N \sum_{k=1}^{M_i^B} \frac{t_{ik}}{\pi^A_{ik}} \frac{Y_i}{M_i^B}. \end{aligned} \tag{3.27}$$

Since each unit j of U^A corresponds to a unit k of a network i of U^B, the double sum over the population U^B can also be written using a sum over the population U^A. Thus,

$$\hat{Y}^B = \sum_{i=1}^N \sum_{k=1}^{M_i^B} \frac{t_{ik}}{\pi^A_{ik}} \frac{Y_i}{M_i^B} = \sum_{j=1}^{M_A} \frac{t_j}{\pi^A_j} \mu_j = \sum_{j=1}^{m^A} \frac{\mu_j}{\pi^A_j}. \tag{3.28}$$

Therefore, we have $\hat{Y}^{ADAP,B} = \hat{Y}^B$. Thompson (1990) proved that adaptive cluster sampling is unbiased. For the precision of estimates, this type of sampling seems to be worthwhile in comparison to conventional sampling when the target population forms clusters. This remark supports the use of the GWSM since this turns out to be a generalisation of adaptive cluster sampling.

3.6 SNOWBALL SAMPLING

In the context where we are looking to survey clusters of individuals by selecting at the start one or many elements of the clusters, Goodman (1961) suggested *snowball sampling*. This is similar to the type of sampling that concerns us, i.e., surveying entire clusters from the target population U^B. However, the sizes of the clusters are not fixed in advance, but rather by the selection parameters.

A snowball sample with τ phases[3] and κ names can be described in the following way. A random sample s of n individuals is selected from a population of size N where each individual k is selected with a probability $\pi_k > 0$. The sampling design used here to select this sample does not matter much in the survey process. At the first phase, each of the n individuals selected in s is asked to name κ names of individuals belonging to the same population. The way in which the names of the individuals are chosen must be specified in the survey process. For example, we can ask an individual to name κ people from his or her immediate family, or κ people of the same nationality. The individuals named by the individuals selected in s, and who are not part of s, form the first phase of the survey. Note that we create here clusters of size $\kappa+1$ that can however be overlapping. At the second phase, we ask each individual from the first phase to in turn name κ individuals. In a similar way, the new individuals named by the individuals from the first phase, and who are neither part of the first phase nor of s, form the second phase of the survey. This process continues until we have completed τ phases.

Goodman (1961) was interested in this type of sampling not to estimate some total of a variable of interest y, but rather to estimate the number of *relationships* between individuals. A mutual relationship (or of type (1,1)) exists when an individual k names an individual k', and vice versa. A relationship of type $(\tau+1, 1)$ exists

---

[3] The sense of the term "phase" used here by Goodman (1961) differs from that commonly used in sampling theory, namely a design where each phase represents a level of sampling where the second-phase units are selected within the units selected at the first phase, and so on.

when an individual k from s named another individual at the first phase of sampling, who then named another individual, and so on until the individual from phase τ names the first individual k. This relationship containing $\tau+1$ individuals is called circular. Goodman (1961) also studied the estimation of the number of relationships of type (τ,κ), that is, relationships where, counting all the κ individuals named by a given individual k from s, all the individuals named by the κ individuals coming from the previous κ individuals, and so on for the τ phases, we have exactly $\tau+\kappa$ individuals. It is again interesting to note that the relationships studied by Goodman (1961) are nothing but clusters.

Snowball sampling can be similar to the survey process studied here with indirect sampling and the GWSM. Recall that the survey constraint associated with indirect sampling is that all units of the clusters selected from the target population U^B must be surveyed. This in fact corresponds to snowball sampling with $\kappa=1$ phase and $\tau=\infty$ names. Indeed, by selecting the sample s^A and by surveying the corresponding units in U^B, we have, so to speak, the selection of units in U^B. Now, the process looking to survey the rest of the individuals of the cluster corresponds to the survey process where we ask each individual k from cluster i to name all the M_i^B individuals contained in the cluster, whatever the number they are.

Snowball sampling can also be similar to the adaptive cluster sampling of Thompson (1990). If we refer to Figure 3.3, we then have snowball sampling with $\kappa=\infty$ phases and four names. Here, we are looking to survey individuals having some characteristic. We then randomly choose a sample of quadrilaterals (i.e., small squares) of individuals and we identify the quadrilaterals where we found individuals having the desired characteristic. In the identified quadrilaterals, we are then going to see the four adjacent quadrilaterals (north, south, east, and west) to identify other individuals having the desired characteristic. We proceed in this manner until we find no more adjacent quadrilaterals having the characteristic. The process of naming individuals in snowball sampling corresponds here to identifying an adjacent quadrilateral having the desired characteristic. We thus "name" all the quadrilaterals during a sufficient number of phases (not specified in advance), until all the "named" quadrilaterals systematically bring us back to quadrilaterals already named.

CHAPTER 4

PROPERTIES

In this chapter, we present properties of the GWSM. We will first show that the GWSM is unbiased for the estimation of the total Y^B of the target population U^B. We will then give the variance formula of the estimator $\hat{Y}^B$, and we will discuss the estimation of this variance. Afterwards, we will show that, in the case where the indirect sampling carried out is conventional cluster sampling, the GWSM gives the same results as the classical theory. We will then deal with the case where we do a census of the population U^A and where we do a census of the target population U^B. We will also look at the use of weighted links. Finally, we will look to improve the estimator $\hat{Y}^B$ by reducing its variance. For this, we will first be using sufficient statistics, and next we will find optimum weights for the links.

4.1 BIAS AND VARIANCE

In order to be able to calculate the *bias* and the *variance* of the estimator $\hat{Y}^B$, we first prove the following Theorem 4.1.

Theorem 4.1: Duality of the form of $\hat{Y}^B$ with respect to U^A and U^B

Let $z_{ik} = Y_i / L_i^B$ *where* $Y_i = \sum_{k=1}^{M_i^B} y_{ik}$ *and* $L_i^B = \sum_{k=1}^{M_i^B} L_{ik}^B$ *for all* $k \in U_i^B$. *The estimator* $\hat{Y}^B$, *given by (2.1), can then also be written under the form*

$$\hat{Y}^B = \sum_{j=1}^{M^A} \frac{t_j}{\pi_j^A} Z_j \tag{4.1}$$

where
$$Z_j = \sum_{i=1}^{N}\sum_{k=1}^{M_i^B} l_{j,ik} z_{ik} \,. \tag{4.2}$$

Proof:

From $\hat{Y}^B = \sum_{i=1}^{n} w_i \sum_{k=1}^{M_i^B} y_{ik} = \sum_{i=1}^{n} w_i Y_i$, we substitute the definition of w_i in $\hat{Y}^B$ to obtain

$$\hat{Y}^B = \sum_{i=1}^{n} Y_i \left(\frac{\sum_{k=1}^{M_i^B} w'_{ik}}{\sum_{k=1}^{M_i^B} L_{ik}^B} \right) = \sum_{i=1}^{n} \frac{Y_i}{L_i^B} \sum_{k=1}^{M_i^B} w'_{ik} \,. \tag{4.3}$$

Let $z_{ik} = Y_i / L_i^B$ for all $ik \in U^B$. Note that this quantity is defined if and only if $L_i^B > 0$ for all clusters i of U^B, that is, if and only if constraint 2.1 is satisfied. From (4.3), we obtain

$$\hat{Y}^B = \sum_{i=1}^{n}\sum_{k=1}^{M_i^B} w'_{ik} z_{ik} \,. \tag{4.4}$$

By replacing w'_{ik} with its definition (2.2), we get

$$\hat{Y}^B = \sum_{i=1}^{n}\sum_{k=1}^{M_i^B} \left(\sum_{j=1}^{M^A} l_{j,ik} \frac{t_j}{\pi_j^A} \right) z_{ik} = \sum_{i=1}^{n}\sum_{k=1}^{M_i^B}\sum_{j=1}^{M^A} l_{j,ik} t_j \frac{z_{ik}}{\pi_j^A} \,. \tag{4.5}$$

Following the survey process, the M_i^B units of cluster i are surveyed if and only if $l_{j,ik} \neq 0$ (there is a link between units j of U^A and ik of U^B) for at least one $k \in U_i^B$, and $t_j = 1$ (unit j of U^A is selected in s^A), or in other words, if and only if $l_{j,ik} t_j \neq 0$ for at least one $k \in U_i^B$. [1] Therefore, cluster i is surveyed if and only if, for all z_{ik} / π_j^A, we have $l_{j,ik} t_j z_{ik} / \pi_j^A \neq 0$ for at least one $k \in U_i^B$; which corresponds to having $\varphi_i = \sum_{k=1}^{M_i^B} \sum_{j=1}^{M^A} l_{j,ik} t_j z_{ik} / \pi_j^A \neq 0$. The n

---

[1] In the present context, since the variable $l_{j,ik}$ is dichotomous, writing $l_{j,ik} \neq 0$ is equivalent to writing $l_{j,ik} = 1$. This condition will be relaxed, particularly in Section 4.5 and Chapter 9 on record linkage, where we will allow a non-negative real value for $l_{j,ik}$.

surveyed clusters thus have $\varphi_i \neq 0$, and the $N-n$ unsurveyed clusters have $\varphi_i = 0$. Thus,

$$\hat{Y}^B = \sum_{i=1}^{n}\sum_{k=1}^{M_i^B}\sum_{j=1}^{M^A} l_{j,ik} t_j \frac{z_{ik}}{\pi_j^A} = \sum_{i=1}^{N}\sum_{k=1}^{M_i^B}\sum_{j=1}^{M^A} l_{j,ik} t_j \frac{z_{ik}}{\pi_j^A} \quad (4.6)$$

and finally,

$$\hat{Y}^B = \sum_{j=1}^{M^A} \frac{t_j}{\pi_j^A} \sum_{i=1}^{N}\sum_{k=1}^{M^B} l_{j,ik} z_{ik} = \sum_{j=1}^{M^A} \frac{t_j}{\pi_j^A} Z_j . \quad (4.7)$$

The estimator $\hat{Y}^B$ can therefore be written as a function of units *ik* from U^B, or as a function of units *j* from U^A. ■

The estimator $\hat{Y}^B$ is in fact only a Horvitz-Thompson estimator where the variable of interest is the variable Z_j. This observation leads us to many results that then become relatively simple to prove. Note that Deville (1998a) obtained estimator (4.1) by using matrix notation.

***Example* 4.1**

As an example, we return to the case illustrated in Figure 2.1.

Units of U^B from Fig. 2.1	***i***	***k***	z_{ik}
1	1	1	$\frac{y_{11}+y_{12}}{2}$
2	1	2	$\frac{y_{11}+y_{12}}{2}$
3	2	1	$\frac{y_{21}+y_{22}+y_{23}}{3}$
4	2	2	$\frac{y_{21}+y_{22}+y_{23}}{3}$
5	2	3	$\frac{y_{21}+y_{22}+y_{23}}{3}$
6	3	1	$\frac{y_{31}+y_{32}}{1}$
7	3	2	$\frac{y_{31}+y_{32}}{1}$

Z_1	$\dfrac{y_{11}+y_{12}}{2}$
Z_2	$\dfrac{y_{11}+y_{12}}{2}+\dfrac{y_{21}+y_{22}+y_{23}}{3}$
Z_3	$2\times\left(\dfrac{y_{21}+y_{22}+y_{23}}{3}\right)$
Z_4	$\dfrac{y_{31}+y_{32}}{1}$

Suppose that we select from U^A unit j=1 and unit j=2. The estimator $\hat{Y}^B$ given by (4.1) is then written:

$$\hat{Y}^B=\frac{1}{\pi_1^A}\left(\frac{y_{11}+y_{12}}{2}\right)+\frac{1}{\pi_2^A}\left(\frac{y_{11}+y_{12}}{2}+\frac{y_{21}+y_{22}+y_{23}}{3}\right)$$

$$=\frac{1}{2}\left[\frac{1}{\pi_1^A}+\frac{1}{\pi_2^A}\right]y_{11}+\frac{1}{2}\left[\frac{1}{\pi_1^A}+\frac{1}{\pi_2^A}\right]y_{12}+\frac{y_{21}}{3\pi_2^A}+\frac{y_{22}}{3\pi_2^A}+\frac{y_{23}}{3\pi_2^A}.$$

We get the same estimator as the illustrative example from Chapter 2 where this was derived from (2.1).

Corollary 4.1: Bias of $\hat{Y}^B$

The estimator $\hat{Y}^B$ is unbiased for estimating Y^B, with respect to the sampling design.

Proof:

We take the expected value of (4.1) with respect to the sampling design

$$E(\hat{Y}^B)=\sum_{j=1}^{M^A}\frac{E(t_j)}{\pi_j^A}Z_j=\sum_{j=1}^{M^A}Z_j=Z \tag{4.8}$$

as $E(t_j)=\pi_j^A$.

It is then sufficient to prove that Z= Y^B. First we have

$$\begin{aligned}Z&=\sum_{j=1}^{M^A}Z_j=\sum_{j=1}^{M^A}\sum_{i=1}^{N}\sum_{k=1}^{M_i^B}l_{j,ik}z_{ik}\\&=\sum_{i=1}^{N}\sum_{k=1}^{M_i^B}z_{ik}\sum_{j=1}^{M^A}l_{j,ik}=\sum_{i=1}^{N}\sum_{k=1}^{M_i^B}z_{ik}L_{ik}^B.\end{aligned} \tag{4.9}$$

Since $z_{ik}=Y_i/L_i^B$, we therefore have

$$Z = \sum_{i=1}^{N}\sum_{k=1}^{M_i^B} z_{ik} L_{ik}^B = \sum_{i=1}^{N}\sum_{k=1}^{M_i^B} \frac{Y_i}{L_i^B} L_{ik}^B$$
$$= \sum_{i=1}^{N} \frac{Y_i}{L_i^B} \sum_{k=1}^{M_i^B} L_{ik}^B = \sum_{i=1}^{N} Y_i = Y^B. \quad (4.10)$$

■

The unbiasedness of the GWSM can also be shown with the help of a similar method as that presented by Ernst (1989).

Corollary 4.2: Variance of $\hat{Y}^B$

The formula for the variance of the estimator $\hat{Y}^B$, with respect to the sampling design, is given by:

$$Var(\hat{Y}^B) = \sum_{j=1}^{M^A}\sum_{j'=1}^{M^A} \frac{(\pi_{jj'}^A - \pi_j^A \pi_{j'}^A)}{\pi_j^A \pi_{j'}^A} Z_j Z_{j'} \quad (4.11a)$$

or, equivalently, by

$$Var(\hat{Y}^B) = -\frac{1}{2}\sum_{j=1}^{M^A}\sum_{j'=1}^{M^A} (\pi_{jj'}^A - \pi_j^A \pi_{j'}^A)\left(\frac{Z_j}{\pi_j^A} - \frac{Z_{j'}}{\pi_{j'}^A}\right)^2, \quad (4.11b)$$

where we denote by $\pi_{jj'}^A$ the joint probability of the selection of units j and j'.

Proof:

To obtain a variance formula for $\hat{Y}^B$, we start from equation (4.1). Since it turns out that $\hat{Y}^B$ is nothing more than a Horvitz-Thompson estimator of the total Z, the variance of $\hat{Y}^B$ follows directly. For details of the proof, see Särndal, Swensson and Wretman (1992). ■

For the calculation of $\pi_{jj'}^A$ under various sampling designs, one can look at Särndal, Swensson and Wretman (1992).

In practice, equations (4.11a) and (4.11b) are easy to set up. It is sufficient at first to calculate $z_{ik} = Y_i / L_i^B$ for each unit k of each surveyed cluster i. We then calculate the total $Z_j = \sum_{i=1}^{N}\sum_{k=1}^{M_i^B} l_{j,ik} z_{ik}$. All that remains is to substitute each Z_j in the variance equation of the Horvitz-Thompson estimator.

The variance $Var(\hat{Y}^B)$ can be estimated without bias using the following equation:

$$\hat{V}ar(\hat{Y}^B) = \sum_{j=1}^{m^A}\sum_{j'=1}^{m^A} \frac{(\pi_{jj'}^A - \pi_j^A \pi_{j'}^A)}{\pi_{jj'}^A \pi_j^A \pi_{j'}^A} Z_j Z_{j'} \tag{4.12}$$

(Särndal, Swensson and Wretman, 1992).

We can also draw up another variance estimator of $Var(\hat{Y}^B)$ inspired from Yates and Grundy (1953). This estimator is given by

$$\hat{\hat{V}}ar(\hat{Y}^B) = -\frac{1}{2}\sum_{j=1}^{m^A}\sum_{j'=1}^{m^A} \frac{(\pi_{jj'}^A - \pi_j^A \pi_{j'}^A)}{\pi_{jj'}^A} \left(\frac{Z_j}{\pi_j^A} - \frac{Z_{j'}}{\pi_{j'}^A} \right)^2 . \tag{4.13}$$

Other variance estimators are proposed in the scientific literature, such as Jackknife and Bootstrap estimators. We will present in Chapter 6 a Jackknife variance estimator used within the context of SLID. For more information, we can consult Wolter (1985) and Särndal, Swensson and Wretman (1992).

4.2 PARTICULAR CASE 1: CLUSTER SAMPLING

We saw that the GWSM allows for the calculation of estimation weights in the case of indirect sampling where the target population U^B consists of clusters. In the context of conventional cluster sampling, the question is then to know if the GWSM gives the same results as classical theory.

Cluster sampling was presented in section 1.2. We recall that this type of sampling consists of first selecting primary sampling units (PSU) that contain secondary sampling units (SSU). Finally, we survey all the SSU belonging to the selected PSU.

In the context of indirect sampling, we can illustrate cluster sampling with the help of Figure 4.1.

Using the notation relative to the GWSM, we select a sample s^A containing m^A PSU in the population U^A containing M^A PSU according to a certain sampling design. We assume that π_j^A represents the selection probability of PSU j, where $\pi_j^A > 0$ for all $j \in U^A$. The target population U^B contains M^B units. This population is divided into N clusters, where cluster i contains M_i^B units. Each cluster i of

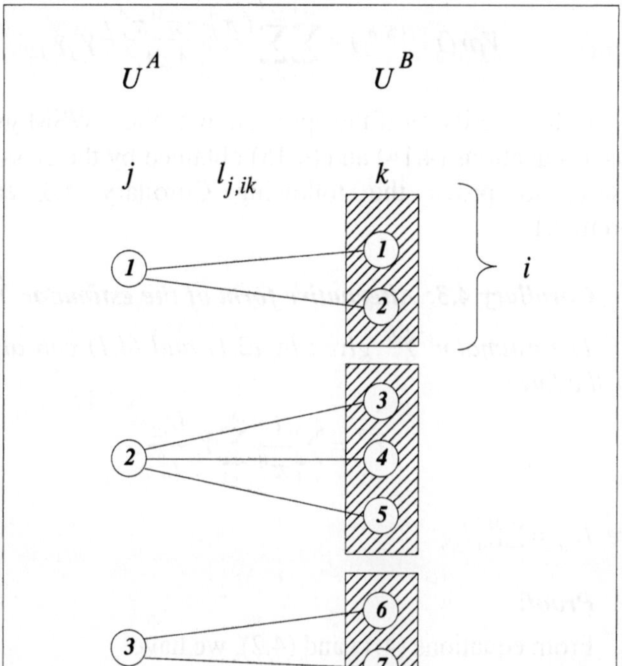

Figure 4.1: *Example of links in cluster sampling*

U^B is linked to a PSU j of U^A. The links l between populations U^A and U^B are thus made between PSU j and SSU k of the clusters i. Note here that the two indices j and i represent the clusters and thus, these two indices are interchangeable.

With cluster sampling, we are looking to estimate the total $Y^B = \sum_{i=1}^{N} \sum_{k=1}^{M_i^B} y_{ik}$ for a characteristic y. To do this, the classical theory suggests using the estimator $\hat{Y}^{CLUS,B}$ given by

$$\hat{Y}^{CLUS,B} = \sum_{j=1}^{m^A} \frac{Y_j}{\pi_j^A} \tag{4.14}$$

where $Y_j = Y_i = \sum_{k=1}^{M_i^B} y_{ik}$ for $j=i$ (Särndal, Swensson and Wretman, 1992). The variance of $\hat{Y}^{CLUS,B}$ is given by

$$Var(\hat{Y}^{CLUS,B}) = \sum_{j=1}^{M^A}\sum_{j'=1}^{M^A} \frac{(\pi_{jj'}^A - \pi_j^A \pi_{j'}^A)}{\pi_j^A \pi_{j'}^A} Y_j Y_{j'} . \qquad (4.15)$$

Before verifying if the application of the GWSM gives the same results as equations (4.14) and (4.15) obtained by the classical theory, it is useful to prove the following Corollary 4.3 ensuing from Theorem 4.1.

Corollary 4.3: Alternative form of the estimator $\hat{Y}^B$

The estimator $\hat{Y}^B$ given by (2.1) and (4.1) can also be written under the form

$$\hat{Y}^B = \sum_{j=1}^{M^A} \frac{t_j}{\pi_j^A} \sum_{i=1}^{N} Y_i \frac{L_{j,i}}{L_i^B}, \qquad (4.16)$$

where $L_{j,i} = \sum_{k=1}^{M_i^B} l_{j,ik}$.

Proof:

From equations (4.1) and (4.2), we have

$$\hat{Y}^B = \sum_{j=1}^{M^A} \frac{t_j}{\pi_j^A} \sum_{i=1}^{N} \sum_{k=1}^{M_i^B} l_{j,ik} z_{ik} . \qquad (4.17)$$

By replacing z_{ik} with its definition, we then get

$$\begin{aligned} \hat{Y}^B &= \sum_{j=1}^{M^A} \frac{t_j}{\pi_j^A} \sum_{i=1}^{N} \sum_{k=1}^{M_i^B} l_{j,ik} \frac{Y_i}{L_i^B} \\ &= \sum_{j=1}^{M^A} \frac{t_j}{\pi_j^A} \sum_{i=1}^{N} \frac{Y_i}{L_i^B} \sum_{k=1}^{M_i^B} l_{j,ik} \qquad (4.18) \\ &= \sum_{j=1}^{M^A} \frac{t_j}{\pi_j^A} \sum_{i=1}^{N} Y_i \frac{L_{j,i}}{L_i^B}. \end{aligned}$$

■

Note that the form of estimator (4.16) reminds the multiplicity approach described in section 3.2 and presented by Huang (1984). In fact, estimator (4.16) is a generalisation of the multiplicity approach.

We can now check if the application of the GWSM gives the same results as equations (4.14) and (4.15) obtained by the classical theory. First of all, since the indices j and i are interchangeable, we have

$$L_{j,i} = \sum_{k=1}^{M_i^B} l_{j,ik} = \begin{cases} L_i^B & \text{if } j = i \\ 0 & \text{otherwise.} \end{cases} \tag{4.19}$$

The ratio $L_{j,i} / L_i^B$ is thus equal to 1 when $j=i$, and 0 otherwise. For a given PSU j, we then have the following result:

$$\sum_{i=1}^{N} Y_i \frac{L_{j,i}}{L_i^B} = Y_i = Y_j. \tag{4.20}$$

From (4.14) and (4.20), we can then rewrite the estimator coming from the GWSM in the following way:

$$\hat{Y}^B = \sum_{j=1}^{M^A} \frac{t_j}{\pi_j^A} Y_j = \sum_{j=1}^{m^A} \frac{Y_j}{\pi_j^A} = \hat{Y}^{CLUS,B}. \tag{4.21}$$

The variance formula (4.15) follows directly from this result.

Thus, in the case of conventional cluster sampling, the GWSM gives the same results as the classical theory. This suggests that in simple estimation cases where the GWSM possibly would not have been essential, the results obtained would be comparable to those coming from a more classical theory.

4.3 PARTICULAR CASE 2: CENSUS OF POPULATION U^A

In Chapter 2, the GWSM was presented in the context where a sample s^A containing m^A units was selected from the population U^A containing M^A units according to a certain sampling design. Using the links between population U^A and the target population U^B, we then looked to estimate the total Y^B using the sample s^A.

We can now ask ourselves what happens to the precision of the estimator $\hat{Y}^B$ if we perform a census of U^A instead of selecting a sample.

In the case of a *census* of U^A, we have $\pi_j^A = 1$ and $t_j = 1$ for all units $j \in U^A$, and thus $s^A = U^A$. From Theorem 4.1, we then have $\hat{Y}^B = \sum_{j=1}^{M^A} Z_j = Z$. From (4.10), we have $Z = Y^B$ and thus, we get directly $\hat{Y}^B = Y^B$.

By performing a census of U^A, the total Y^B can then be estimated with certainty. Note that the inverse reasoning is not always true. Indeed, if we perform a census of the population U^B, the estimator $\hat{Y}^B$ will not necessarily have zero variance. This situation is discussed in the following section.

We will also see in section 5.3 that in setting up for U^B a single cluster of size M^B (which, following the survey process, will be completely surveyed), the estimator $\hat{Y}^B$ in general does not have a variance equal to zero.

4.4 PARTICULAR CASE 3: CENSUS OF POPULATION U^B

With indirect sampling, it is possible for the selection of certain samples s^A of U^A to lead to a census of the target population U^B. If this occurs for a subset of all the possible samples selected from U^A, we cannot expect to get zero variance for the estimator $\hat{Y}^B$.

Unfortunately, such is also the case even if all the possible samples s^A of U^A lead to a census of U^B. In other words, if the number of surveyed clusters n corresponds to the total number N of clusters from U^B, the variance of $\hat{Y}^B$ is not necessarily zero.

The small example that we present in Figure 4.2 on the census of population U^B, perfectly illustrates this situation.

Let there exist populations U^A and U^B represented in Figure 4.2. The population U^A contains three units with two units selected using simple random sampling without replacement. The target population U^B contains two clusters of size 1. Using the links between the population U^A and the target population U^B, we look to estimate the total Y^B from the sample s^A using the estimator $\hat{Y}^B$ given by (4.1). To use this estimator, we must first calculate the values of the derived variables Z_j given by (4.2).

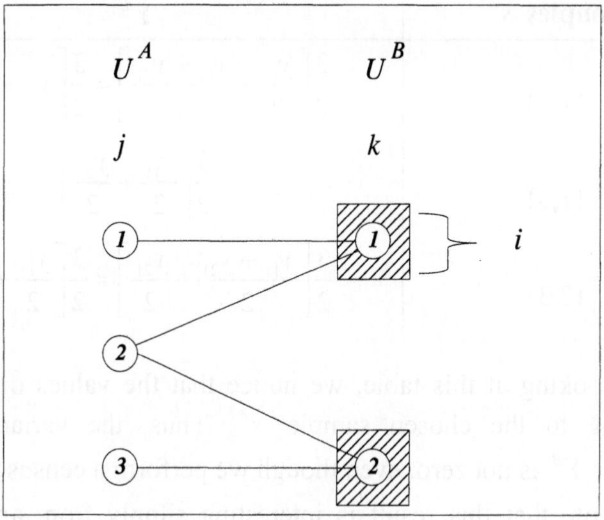

Figure 4.2: *Example of census of U^B*

Z_1	$\frac{y_{11}}{2}$
Z_2	$\frac{y_{11}}{2}+\frac{y_{21}}{2}=\frac{y_{11}+y_{21}}{2}$
Z_3	$\frac{y_{21}}{2}$

There are three possible samples s^A:

$$\{1,2\},\ \{2,3\},\text{ and }\{1,3\}.$$

For each sample, we always survey the two clusters of U^B. Thus, $n=N$ and we then have a census of the target population U^B for all possible samples selected from U^A. For each possible sample s^A, we now calculate the value of the estimator $\hat{Y}^B$, which, in this example, leads to the results contained in the following table.

Samples s^A	$\hat{Y}^B$
{1,2}	$\frac{3}{2}\left[\frac{y_{11}}{2}+\frac{y_{11}+y_{21}}{2}\right]=\frac{3}{2}\left[y_{11}+\frac{y_{21}}{2}\right]$
{1,3}	$\frac{3}{2}\left[\frac{y_{11}}{2}+\frac{y_{21}}{2}\right]$
{2,3}	$\frac{3}{2}\left[\frac{y_{11}+y_{21}}{2}+\frac{y_{21}}{2}\right]=\frac{3}{2}\left[\frac{y_{11}}{2}+y_{21}\right]$

Looking at this table, we notice that the values of $\hat{Y}^B$ differ according to the chosen sample s^A. Thus, the variance of the estimator $\hat{Y}^B$ is not zero, even though we perform a census of U^B.

Note that this result is interesting simply from an academic point of view. In practice, immediately after noticing that the variable y is measured for all units k of the N clusters i of the target population U^B, the estimator $\hat{Y}^B$ could be directly replaced by the measured value of the total Y^B. We would no longer be faced with indirect sampling, but rather a direct census of U^B.

4.5 USE OF WEIGHTED LINKS

With indirect sampling, we assume that a link (or a relationship) existed between units j of population U^A and units ik of population U^B. In chapter 2, this link is identified by an indicator variable $l_{j,ik}$, where $l_{j,ik}=1$ if a link exists between unit $j \in U^A$ and unit $ik \in U^B$, and 0 otherwise. The variable $l_{j,ik}$ simply indicates that there is or not a link between units j and ik from populations U^A and U^B. It does not, however, indicate the relative importance that certain links can have compared to others.

Take the case of a survey of enterprises where we have a unit j from the sampling frame U^A that is linked to two establishments k=1 and k=2 of enterprise i of the target population U^B. Suppose that establishment k=1 has 1 million euros in assets and establishment k=2 100 million euros. In the construction of an economic indicator, we could then want to give a larger weight to establishment k=2 compared to establishment k=1. In the context of the GWSM, this could happen by replacing the indicator variable $l_{j,ik}$ with a quantitative variable $\theta_{j,ik}$ representing the assets of the establishments.

In a more general context, it is possible to replace the indicator variable $l_{j,ik}$ with any quantitative variable $\theta_{j,ik}$ representing the importance that we want to give to the link $l_{j,ik}$. It indeed turns out that there is no problem with generalising the indicator variable l defined on $\{0,1\}$ with a quantitative variable θ defined on $[0,+\infty[$, the set of non-negative real numbers. The theory developed around the GWSM remains quite valid.

Remember that a value of $\theta_{j,ik} = 0$ for two units j of U^A and ik of U^B amounts to a link $l_{j,ik} = 0$. In order for the GWSM to remain unbiased, it is always necessary to respect the following constraint.

Constraint 4.1

For each cluster i of U^B, we must have

$$\theta_i^B = \sum_{j=1}^{M^A} \sum_{k=1}^{M_i^B} \theta_{j,ik} > 0.$$

With the use of a quantitative variable $\theta_{j,ik}$ instead of the indicator variable $l_{j,ik}$, we can explain why the PCF described in section 3.1 appears in the possible applications of the GWSM. Recall that the PCF is a factor associated with variables measuring partners and which decreases the value of these variables proportionally to the profits of the partners owning the enterprise. Let us go back to the example of the enterprise that has two partners, where each partner earns 50% of the profits of the enterprise. This corresponds, let's say, to having a cluster i=1 (the enterprise) of U^B having only a single unit k=1 and where this unit k=1 is linked to two units j=1 and j=2 (the partners) of U^A.

Now, consider the estimator $\hat{Y}^B$ under the form given by Theorem 4.1. From (4.2), for j=1, 2, the value of Z_j is given here by $Z_j = Y_1 / 2$. If we replace in (4.2) the indicator variable $l_{j,ik}$ by a quantitative variable $\theta_{j,ik}$ representing the profits of partner j in establishment k=1 of enterprise i=1, the variable Z_j is then given by $Z_j = \theta_{j,1} Y_1 / \theta_1^B$ where $\theta_{j,1} = \theta_{j,11}$ since enterprise i=1 only has a single establishment k=1, and where $\theta_1^B = \theta_{1,11} + \theta_{2,11}$. Here, the PCF is given by $\theta_{j,1} / \theta_1^B$ and it effectively decreases the value of the variable of interest Y_1 measured for the enterprise, proportionally to the profits of the partners owning the enterprise.

Setting $\tilde{\theta}_{j,ik} = \theta_{j,ik} / \theta_i^B$ where $\theta_i^B = \sum_{j=1}^{M^A} \sum_{k=1}^{M_i^B} \theta_{j,ik}$, we get a direct generalisation to the WSM described in section 3.3. Note that we then have $\sum_{j=1}^{M^A} \sum_{k=1}^{M_i^B} \tilde{\theta}_{j,ik} = 1$. This generalisation leads to a version of the GWSM constructed with the constants $\tilde{\theta}_{j,ik}$.

Steps for the GWSM with weighted links

Step 1: For each unit k of clusters i of Ω^B, we calculate the initial weight $w_{ik}'^{\theta}$, that is:

$$w_{ik}'^{\theta} = \sum_{j=1}^{M^A} \tilde{\theta}_{j,ik} \frac{t_j}{\pi_j^A},$$

where $t_j = 1$ if $j \in s^A$, and 0 otherwise.

Step 2: The final weight w_i^{θ} is given by

$$w_i^{\theta} = \sum_{k=1}^{M_i^B} w_{ik}'^{\theta} = \sum_{k=1}^{M_i^B} \sum_{j=1}^{M^A} \tilde{\theta}_{j,ik} \frac{t_j}{\pi_j^A}.$$

Step 3: Finally, we set $w_{ik}^{\theta} = w_i^{\theta}$ for all $k \in U_i^B$.

In section 4.6.3, we will seek for optimal values for the weighted links $\theta_{j,ik}$ (or $\tilde{\theta}_{j,ik}$). In Chapter 9, we will go into greater depth on the use of the quantitative variable θ, particularly in the case where θ is the linkage weight coming from a probabilistic record linkage.

4.6 IMPROVEMENT OF THE ESTIMATOR

The GWSM, as we have seen, offers a simple solution of obtaining an estimation weight w_{ik} for each unit k of the surveyed clusters i. However, the resulting estimator $\hat{Y}^B$ given by (2.1) is not always the one that has the smallest variance.

It is in fact possible to improve the estimator $\hat{Y}^B$ using a conditional approach, or sufficient statistics. It is also possible to improve it by determining optimal weights for the links $\theta_{j,ik}$ presented in section 4.5. The estimators obtained by these approaches have a variance less than or equal to that of the original estimator $\hat{Y}^B$.

4.6.1 Conditional approach

The *conditional approach* consists of improving the estimator $\hat{Y}^B$ by obtaining a new estimator $\hat{Y}^{COND,B}$ based on the conditional expectation of $\hat{Y}^B$, given a certain statistic ϑ. The new estimator $\hat{Y}^{COND,B}$ is thus defined from

$$\hat{Y}^{COND,B} = E(\hat{Y}^B \mid \vartheta). \tag{4.22}$$

The conditional approach is in fact based on the following identity:

$$Var(\hat{Y}^B) = Var_\vartheta[E(\hat{Y}^B \mid \vartheta)] + E_\vartheta[Var(\hat{Y}^B \mid \vartheta)] \tag{4.23}$$

where $E_\vartheta(.)$ and $Var_\vartheta(.)$ are calculated with respect to all possible values of the statistic ϑ. We note that $E_\vartheta[Var(\hat{Y}^B \mid \vartheta)] \geq 0$ and therefore $Var(\hat{Y}^{COND,B}) = Var_\vartheta[E(\hat{Y}^B \mid \vartheta)] \leq Var(\hat{Y}^B)$. The variance of the estimator $\hat{Y}^{COND,B}$ is thus less than or equal to the variance of the estimator $\hat{Y}^B$.

With the GWSM, it proves to be useful to condition on the set Ω^B of n clusters identified by the units j of the sample s^A. Recall that we perform the selection of the sample s^A of m^A units in the population U^A. For each unit j selected in s^A, we then identify the units ik of U^B that have a non-zero link with j, i.e., $l_{j,ik} = 1$. For each unit ik identified, we measure the variable y_{ik} for all M_i^B units of the cluster i containing this unit. By looking at the sampling design from the point of view of the population U^B, we can see this as cluster sampling where we obtained a sample Ω^B of n clusters selected among the N clusters of population U^B. The selection of each unit j of s^A therefore leads to the selection of a cluster i from U^B and we notice that there can be many samples s^A leading to a set Ω^B of given clusters.

Starting from estimator (4.16), the new estimator $\hat{Y}^{COND,B}$ is given by

$$\begin{aligned}\hat{Y}^{COND,B} &= E(\hat{Y}^B \mid \Omega^B) = E\left[\sum_{j=1}^{M^A} \frac{t_j}{\pi_j^A} \sum_{i=1}^{N} Y_i \frac{L_{j,i}}{L_i^B} \,\middle|\, \Omega^B\right] \\ &= \sum_{j=1}^{M^A} \frac{E(t_j \mid \Omega^B)}{\pi_j^A} \sum_{i=1}^{N} Y_i \frac{L_{j,i}}{L_i^B}.\end{aligned} \tag{4.24}$$

Since we have

$$E(t_j \mid \Omega^B) = 1 \times P(t_j = 1 \mid \Omega^B) + 0 \times P(t_j = 0 \mid \Omega^B) = P(t_j = 1 \mid \Omega^B),$$

we can consequently write:

$$\begin{aligned}\hat{Y}^{COND,B} &= \sum_{j=1}^{M^A} \frac{P(t_j = 1 \mid \Omega^B)}{\pi_j^A} \sum_{i=1}^{N} Y_i \frac{L_{j,i}}{L_i^B} \\ &= \sum_{j=1}^{M^A} \sum_{i=1}^{N} \frac{P(t_j = 1 \mid \Omega^B)}{\pi_j^A} L_{j,i} \frac{Y_i}{L_i^B}.\end{aligned} \tag{4.25}$$

The probability $P(t_j = 1 \mid \Omega^B)$ in fact corresponds to the probability of having selected unit j from U^A, considering that the n clusters i of Ω^B had been surveyed. We note that, for a given unit j, if $L_{j,i} = 0$ for all the n clusters $i \in \Omega^B$, we must have $P(t_j = 1 \mid \Omega^B) = 0$. Still for a given unit j, for the $N-n$ clusters $i \notin \Omega^B$, we also have $P(t_j = 1 \mid \Omega^B) = 0$. Indeed, by way of the survey process, the selection of each unit j must lead to the survey of clusters forming Ω^B. Therefore, we get

$$\begin{aligned}\hat{Y}^{COND,B} &= \sum_{j=1}^{M^A} \sum_{i=1}^{n} \frac{P(t_j = 1 \mid \Omega^B)}{\pi_j^A} L_{j,i} \frac{Y_i}{L_i^B} \\ &= \sum_{i=1}^{n} \frac{Y_i}{L_i^B} \sum_{j=1}^{M^A} \frac{P(t_j = 1 \mid \Omega^B)}{\pi_j^A} L_{j,i}.\end{aligned} \tag{4.26}$$

The probabilities $P(t_j = 1 | \Omega^B) \neq 0$ depend upon the links $l_{j,ik} = 1$ linking unit *j* of U^A to the units *k* from clusters *i* of U^B. With complex links, this probability is difficult to set up. Nevertheless, since $P(t_j = 1 | \Omega^B) = P(t_j = 1 , \Omega^B) / P(\Omega^B)$ and $P(\Omega^B)$ is the probability of surveying the *n* clusters of Ω^B, it is clear that it is a function of the selection probabilities of **all** the units *j* having non-zero links with these *n* clusters of Ω^B.

Unfortunately, although through identity (4.23) the estimator $\hat{Y}^{COND,B}$ has a variance smaller than or equal to that of $\hat{Y}^B$, the estimator $\hat{Y}^{COND,B}$ here only has a theoretical interest. Indeed, we saw in section 2.2 that one of the major uses of the GWSM is to be able to get a weighting using only the selection probabilities π_j^A of the units selected in s^A. Unfortunately, $P(t_j = 1 | \Omega^B)$ is a function of the selection probabilities of all units *j* having non-zero links with the *n* clusters of Ω^B, whether these units have been selected or not. As already mentioned, there exist many situations where the probabilities π_j^A are unknown for the units $j \notin s^A$. In these cases, it is then impossible to use the estimator $\hat{Y}^{COND,B}$, and the only estimator $\hat{Y}^B$ obtained by the GWSM remains as one of the sole recourses.

Example 4.2

Suppose that the population U^B has two clusters where each of the clusters has only one unit. This is illustrated in Figure 4.3. Unit *k*=1 of cluster *i*=1 is linked to two units *j*=1 and *j*=2 of U^A, and unit *k*=1 of cluster *i*=2 is linked to a single unit *j*=3 of U^A. We select a sample s^A of size 1. Suppose that we selected unit *j*=2. Then, we survey cluster *i*=1 in order to measure the variable of interest y_{ik} for *i*=1 and *k*=1.

The probability $P(\Omega^B)$ here comes down to the probability of surveying cluster *i*=1 of Ω^B. This probability is equal to the probability of selecting unit *j*=1, unit *j*=2 or the two units *j*=1 and *j*=2 of U^A. Therefore, we have $P(\Omega^B) = \pi_1^A + \pi_2^A - \pi_{12}^A$ where π_{12}^A is the selection probability of the two units *j*=1 and *j*=2. On the other hand, for this example, $P(t_j = 1 , \Omega^B) = P(t_j = 1) = \pi_j^A$ for *j*=1 and *j*=2, and $P(t_3 = 1 , \Omega^B) = 0$ as cluster *i*=2 is not part of Ω^B. Therefore, we

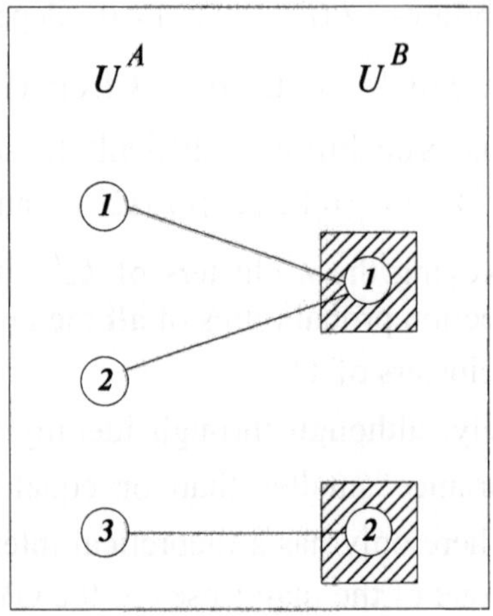

Figure 4.3: *Example of populations for the improvement of estimator* $\hat{Y}^B$

ultimately get $P(t_j = 1 \mid \Omega^B) = \pi_j^A / (\pi_1^A + \pi_2^A - \pi_{12}^A)$ for j=1 and j=2, and $P(t_3 = 1 \mid \Omega^B) = 0$.

From (4.16), the estimator $\hat{Y}^B$ obtaining by the GWSM is here given by $\hat{Y}^B = \dfrac{Y_1}{2\pi_2^A}$.

On the other hand, the estimator (4.26) obtained by the conditional approach is given by

$$\hat{Y}^{COND,B} = \frac{Y_1}{2}\sum_{j=1}^{2}\frac{1}{(\pi_i^A + \pi_2^A - \pi_{12}^A)} = \frac{Y_1}{(\pi_i^A + \pi_2^A - \pi_{12}^A)}.$$

By comparing these two estimators, we notice that the estimator $\hat{Y}^{COND,B}$ requires the knowledge of the selection probabilities π_1^A, π_2^A and π_{12}^A, while the estimator $\hat{Y}^B$ only requires us to know the selection probability of unit j=2 that was selected in s^A. It is worth noting that since $P(\Omega^B) = \pi_1^A + \pi_2^A - \pi_{12}^A$, the estimator $\hat{Y}^{COND,B}$ here corresponds to the Horvitz-Thompson estimator.

4.6.2 Use of sufficient statistics

Sufficient statistics play a key role in mathematical statistics. We will verify this in the present subsection. By the *Rao-Blackwell theorem*, sufficient statistics allow for the improvement of an existing estimator by producing a new estimator whose mean squared error is less than or equal to that of the starting estimator. Note that this form of improvement was used, among others, by Thompson (1990) in the context of adaptive cluster sampling.

The theory presented here on sufficient statistics comes primarily from Cassel, Särndal and Wretman (1977). Note that Thompson and Seber (1996) also gave a similar presentation.

Let $\mathbf{Y}^T = (y_1,\ldots,y_N)$ be the vector containing the values y_k for a population of size *N*. Recall that a sample design **p** is a function $\mathbf{p}(s)$ on the set Ξ of all samples s such that $\mathbf{p}(s) \geq 0$ and $\sum_{s\in\Xi}\mathbf{p}(s) = 1$. We define $\mathbf{D} = \{(k, y_k) \mid k \in s\}$, the set of indices k and the measured variables y_k for the sample *s*.

A statistic $u(\mathbf{D})$ is called *sufficient* for the parameter **Y** if and only if the conditional distribution of **D**, given $u(\mathbf{D})$, does not depend on **Y**, provided that this conditional probability is well-defined.

The statistic $u(\mathbf{D})$ is in fact sufficient if and only if we have the following result:

$$\mathbf{p}(\mathbf{D},\mathbf{Y}) = \mathbf{p}(u(\mathbf{D}),\mathbf{Y}) \times \mathbf{h}(\mathbf{D}) \tag{4.27}$$

where $\mathbf{h}(\mathbf{D}) > 0$ does not depend on **Y**. For the proof of this result, we can consult Bickel and Doksum (1977).

We now present a version of the Rao-Blackwell theorem adapted for finite populations. The Rao-Blackwell theorem was developed independently by Rao (1945) and Blackwell (1947). The first use of this theorem in the context of finite populations was in Basu (1958).

Theorem 4.2 (Rao-Blackwell)

*Let $\hat{Y} = \hat{Y}(\mathbf{D})$ be an estimator (not necessarily unbiased) of $Y = \sum_{k=1}^{N} y_k$ based on the set **D**, and let $u(\mathbf{D})$ be a sufficient statistic.*

We define a new estimator $\hat{Y}^{RB}$ *given by the expectation of* $\hat{Y}$ *conditional on* $u(\mathbf{D})$, *i.e.,*

$$\hat{Y}^{RB} = E(\hat{Y} \mid u(\mathbf{D})) \tag{4.28}$$

We then have,

(a) $E(\hat{Y}^{RB}) = E(\hat{Y})$

(b) $EQM(\hat{Y}) = EQM(\hat{Y}^{RB}) + E[(\hat{Y} - \hat{Y}^{RB})^2]$

(c) $EQM(\hat{Y}^{RB}) \leq EQM(\hat{Y})$.

For the proof of this theorem, we can consult Cassel, Särndal and Wretman (1977).

Let $\mathbf{p}(s^A \mid \mathbf{Y}^B)$ be the sample design associated with the selection of the sample s^A for the measurement of certain values of the vector $\mathbf{Y}^{T,B} = (y_1,...,y_N)$ for the target population U^B. Let $\mathbf{D}_j = \{(i, Y_i) \mid L_{j,i} \neq 0\}$ be the set of indices i and the measured variables Y_i from the clusters of U^B that have at least one link with unit j of U^A. We note that the sets $\mathbf{D}_j$ are not exclusive for $j = 1,..., M^A$. By Corollary 4.3, we notice that the sampling of each unit j from s^A leads to the selection of each cluster i of U^B that have $L_{j,i} \neq 0$. We thus define $\mathbf{D}^A = \{(j, \mathbf{D}_j) \mid j \in s^A\}$ to be the set of the indices j of s^A and the values Y_i of the clusters surveyed through each $j \in s^A$. The set of surveyed clusters Ω^B is thus a function of the sample s^A. Now, let $\mathbf{D}^B = \{(i, Y_i) \mid j \in \Omega^B\}$ be the set of indices i and the measured variables Y_i from clusters $i \in \Omega^B$. As the sampling of each unit j of s^A leads to the selection of at least one cluster i of U^B, the set $\mathbf{D}^B$ is a function of the set $\mathbf{D}^A$, that is $\mathbf{D}^B = u(\mathbf{D}^A)$. Furthermore, we have $\mathbf{p}(\mathbf{D}^A \mid \mathbf{Y}^B) = \mathbf{p}(\mathbf{D}^A, \mathbf{D}^B \mid \mathbf{Y}^B)$.

Using the conditional probabilities, we obtain

$$\mathbf{p}(\mathbf{D}^A, \mathbf{D}^B \mid \mathbf{Y}^B) = \mathbf{p}(\mathbf{D}^A \mid \mathbf{D}^B, \mathbf{Y}^B)\mathbf{p}(\mathbf{D}^B \mid \mathbf{Y}^B).$$

Because the set of surveyed clusters Ω^B is a function of the sample s^A, the selection of sample s^A, given by the set Ω^B, does not depend upon $\mathbf{Y}^{T,B} = (y_1,...,y_N)$. Consequently, we get $\mathbf{p}(\mathbf{D}^A \mid \mathbf{D}^B, \mathbf{Y}^B) = \mathbf{p}(\mathbf{D}^A \mid \mathbf{D}^B)$. We thus finally get the following result:

$$\begin{aligned} \mathbf{p}(\mathbf{D}^A \mid \mathbf{Y}^B) &= \mathbf{p}(\mathbf{D}^B \mid \mathbf{Y}^B) \times \mathbf{p}(\mathbf{D}^A \mid \mathbf{D}^B) \\ &= \mathbf{p}(u(\mathbf{D}^A) \mid \mathbf{Y}^B) \times \mathbf{p}(\mathbf{D}^A \mid \mathbf{D}^B) \end{aligned} \tag{4.29}$$

By comparing (4.27) and (4.29), we then get that the set $\mathbf{D}^B = \{(i, Y_i) \mid j \in \Omega^B\}$ of indices i and measured variables Y_i from clusters $i \in \Omega^B$ is a sufficient statistic for $\mathbf{Y}^B$.

From Corollary 4.3, we have $\hat{Y}^B = \hat{Y}^B(\mathbf{D}^A)$. By Theorem 4.2 (Rao-Blackwell), we can then get a new estimator $\hat{Y}^{RB,B}$ with the sufficient statistic $\mathbf{D}^B = u(\mathbf{D}^A)$ whose mean squared error will be less than or equal to that of $\hat{Y}^B$. Using expression (4.28), this estimator is given by

$$\begin{aligned} \hat{Y}^{RB,B} &= E(\hat{Y}^B \mid u(\mathbf{D}^A)) \\ &= E(\hat{Y}^B \mid \mathbf{D}^B). \end{aligned} \tag{4.30}$$

Because the measurement of Y_i for each cluster i is directly related to Ω^B, we have $E(\hat{Y}^B \mid \mathbf{D}^B) = E(\hat{Y}^B \mid \Omega^B)$. By comparing (4.24) and (4.30), we see that the estimator $\hat{Y}^{RB,B}$ here is the same as the estimator $\hat{Y}^{COND,B}$ obtained by the conditional approach.

Once again, although through Theorem 4.2 (Rao-Blackwell) the estimator $\hat{Y}^{RB,B}$ has a mean squared error less than or equal to that of $\hat{Y}^B$, the estimator $\hat{Y}^{RB,B}$ here only has a theoretical interest. It is indeed a function of selection probabilities for all units j having non-zero links with the n clusters of Ω^B, whether these units were selected or not. This is contrary to one of the uses of the GWSM, which is to be able to get a weighting using only the selection probabilities π_j^A of the units selected in s^A.

4.6.3 Obtaining optimal weighted links

As mentioned before, it is possible to improve $\hat{Y}^B$ by determining optimal weights for the links presented in section 4.5. This problem has been solved by Lavallée and Deville (2002). The goal is to obtain an estimator that has a variance less than or equal to that of the original estimator $\hat{Y}^B$.

In the present section, we will assume that each cluster i of U^B contains only one unit. This is done without loss of generality since following (2.4), the weights w_{ik} are first computed at the cluster level i. As well, the fact that a unit j of U^A can have a link with more than one unit k in the same cluster i can directly be handled by making the weighted links $\theta_{j,i}$ proportional to the original number of links $L_{j,i} = \sum_{k=1}^{M_i^B} l_{j,ik}$. For example, considering Figure 2.1, we see that unit j=2 leads to unit k=4 of U^B, and that unit j=3 leads to units k=3 and k=4. Therefore, we have $L_i^B = 3$ for the identified cluster, rather than $L_i^B = 2$ that we would have if this cluster had contained only one consolidated unit (i.e., adding together units k=3, k=4 and k=5). If one go with one consolidated unit per cluster, we can then make the weighted link $\theta_{j,i}$ proportional to 2 for j=3, and $\theta_{j,i}$ proportional to 1 for j=2.

For obtaining optimal weighted links, it is convenient to use matrix notation, as done by Lavallée and Deville (2002). Let the correspondence between the two populations U^A and U^B be represented by the *link matrix* $\boldsymbol{\Theta}^{AB} = [\theta_{j,i}^{AB}]$ of size $N^A \times N^B$ where each element $\theta_{j,i}^{AB}$ is greater than or equal to zero.[2] That is, unit j of U^A is related to unit i of U^B provided that $\theta_{j,i}^{AB}$>0, otherwise the two units are not related to each other. For the example of Figure 1.2, the link matrix is given by

$$\boldsymbol{\Theta}^{AB} = \begin{bmatrix} \theta_{1,1}^{AB} & \theta_{1,2}^{AB} & 0 & 0 & 0 & 0 & 0 & 0 \\ \theta_{2,1}^{AB} & \theta_{2,2}^{AB} & 0 & 0 & 0 & 0 & 0 & 0 \\ 0 & 0 & \theta_{3,3}^{AB} & \theta_{3,4}^{AB} & 0 & 0 & 0 & 0 \\ 0 & 0 & \theta_{4,3}^{AB} & \theta_{4,4}^{AB} & \theta_{4,5}^{AB} & 0 & 0 & 0 \\ 0 & 0 & 0 & 0 & \theta_{5,5}^{AB} & \theta_{5,6}^{AB} & 0 & 0 \\ 0 & 0 & 0 & 0 & 0 & 0 & \theta_{6,7}^{AB} & \theta_{6,8}^{AB} \end{bmatrix}$$

[2] For the present section, we add the superscript "*AB*" to the weighted links as we will need to differentiate between links from three populations: U^A, U^B, and U^G.

Obtaining the link matrix $\Theta^{AB} = [\theta_{j,i}^{AB}]$ is a critical issue in indirect sampling. It influences the precision of the estimates. Now, as we saw before, in several applications, the values of $\theta_{j,i}^{AB}$>0 for the linked units are simply set to 1. Since the choice of $\theta_{j,i}^{AB}$>0 for two linked units j and i can affect the precision of the estimates, it is natural to seek for those $\theta_{j,i}^{AB}$ values that will minimise the variance of the estimates. This optimisation problem is solved in the present section.

In matrix notation, the total Y^B of the target population U^B is written as $Y^B = \mathbf{1}^{T,B}\mathbf{Y}^B$ where $\mathbf{1}^B$ is the column vector of 1's of size N^B.[3] Setting $\tilde{\theta}_{j,ik} = \theta_{j,ik} / \theta_i^B$ as in section 4.5, we have $\mathbf{1}^A\mathbf{\Theta}^{AB} = \{\theta_1^{AB},...,\theta_{N^B}^{AB}\}$. We then define the *standardised link matrix* $\tilde{\mathbf{\Theta}}^{AB} = \mathbf{\Theta}^{AB}\ [diag(\mathbf{1}^{T,A}\mathbf{\Theta}^{AB})]^{-1}$, where $diag(\mathbf{v})$ is the square matrix obtained by putting the elements of the row-vector (or column-vector) $\mathbf{v}$ in the diagonal, and 0 elsewhere. Note that in order for the matrix $\tilde{\mathbf{\Theta}}^{AB}$ to be well defined, $[diag(\mathbf{1}^{T,A}\mathbf{\Theta}^{AB})]^{-1}$ must exist, which is the case if and only if $\theta_i^{AB} > 0$ for all $i = 1,\ldots,N^B$. Note that this corresponds to Constraint 2.1.

Theorem 4.3

The link matrix $\tilde{\mathbf{\Theta}}^{AB}$ *is a standardised link matrix if and only if*

$$\tilde{\mathbf{\Theta}}^{T,AB}\mathbf{1}^A = \mathbf{1}^B. \tag{4.31}$$

Proof:

By definition, $\tilde{\mathbf{\Theta}}^{T,AB}\mathbf{1}^A = [diag(\mathbf{1}^{T,A}\mathbf{\Theta}^{AB})]^{-1}\mathbf{\Theta}^{T,AB}\mathbf{1}^A$ is a column vector of size N^B. Now, $\mathbf{\Theta}^{T,AB}\mathbf{1}^A = \mathbf{\theta}_+^{AB} = \{\theta_1^{AB},...,\theta_{N^B}^{AB}\}$. Let b_i be the ith element of $\mathbf{\Theta}^{T,AB}\mathbf{1}^A$ obtained by the product of line i of the matrix $diag(\mathbf{\theta}_+^{AB})^{-1}$ and the vector $\mathbf{\theta}_+^{AB}$. We have $b_i = 0 \times \theta_1^{AB} + ... + (\theta_i^{AB})^{-1} \times \theta_i^{AB} + ... + 0 \times \theta_{N^A}^{AB} = 1$.

---

[3] Note that we use for simplification the notation $\mathbf{1}^B$ instead of $\mathbf{1}^{N^B}$.

Therefore, $\widetilde{\mathbf{\Theta}}^{T,AB}\mathbf{1}^A = \{b_1,...,b_{N^B}\}' = \mathbf{1}^B$. ■

Using Theorem 4.3, we directly obtain Corollary 4.4 that can also be found in Deville (1998a):

Corollary 4.4

$$\begin{aligned} Y^B &= \mathbf{1}^{T,B}\mathbf{Y}^B \\ &= \mathbf{1}^{T,A}\widetilde{\mathbf{\Theta}}^{AB}\mathbf{Y}^B = \sum_{j=1}^{N^A}\sum_{i=1}^{N^B}\frac{\theta_{ji}^{AB}}{\theta_i^{AB}}y_i \end{aligned} \tag{4.32}$$

Let us define the column vector $\mathbf{Z} = \widetilde{\mathbf{\Theta}}^{AB}\mathbf{Y}^B$ of size N^A. Considering each line of **Z**, the variable $z_j = \sum_{i=1}^{N^B}\widetilde{\theta}_{ji}^{AB}y_i$ is defined for each unit j of the population U^A and measured for each unit $j \in s^A$.

Let $\mathbf{W}^T = \{w_1,..., w_{N^B}\}$ where w_i is the estimation weight of unit i of Ω^B, with $w_i = 0$ for $i \notin \Omega^B$. For estimating Y^B, the estimator (2.1) can be rewritten as

$$\hat{Y}^B = \mathbf{W}^T\mathbf{Y}^B. \tag{4.33}$$

In matrix notation, the GWSM can be formulated as follows. Let $\boldsymbol{\pi}^A = \{\pi_1^A,...,\pi_{N^A}^A\}'$ and let $\mathbf{\Pi}^A = diag(\boldsymbol{\pi}^A)$ be the diagonal matrix of size $N^A \times N^A$ containing the selection probabilities used for the selection of sample s^A. Accordingly, let $\mathbf{t}^A = \{t_1^A,...,t_{N^A}^A\}'$ where $t_j^A = 1$ if $j \in s^A$, and 0 otherwise. Let $\mathbf{T}^A = diag(\mathbf{t}^A)$ be the diagonal matrix of size $N^A \times N^A$ containing the indicator variables t_j^A. Starting from $Y^B = \mathbf{1}^{T,A}\widetilde{\mathbf{\Theta}}^{AB}\mathbf{Y}^B = \mathbf{1}^{T,A}\mathbf{Z}$, the estimator (4.1) translates to

$$\hat{Y}^B = \mathbf{1}^{T,A}\mathbf{T}^A(\mathbf{\Pi}^A)^{-1}\mathbf{Z} \tag{4.34}$$

Using the fact that $\mathbf{Z} = \widetilde{\mathbf{\Theta}}^{AB}\mathbf{Y}^B$, we have $\hat{Y}^B = \mathbf{1}^{T,A}\mathbf{T}^A(\mathbf{\Pi}^A)^{-1}\widetilde{\mathbf{\Theta}}^{AB}\mathbf{Y}^B$ and therefore we can define the column vector **W** of weights obtained by the GWSM as

$$\mathbf{W} = \widetilde{\mathbf{\Theta}}^{T,AB}\mathbf{T}^A(\mathbf{\Pi}^A)^{-1}\mathbf{1}^A. \tag{4.35}$$

The vector $\mathbf{W}$ is of size N^B and for each $i = 1,\ldots, N^B$, we have $w_i = \sum_{j=1}^{N^A} t_j^A \tilde{\theta}_{j,i}^{AB} / \pi_j^A$.

By construction, because the estimator (4.34) is a Horvitz-Thompson estimator, the GWSM produces unbiased estimates. We can, in addition, have the following theorem:

Theorem 4.4

The vector of weights $\mathbf{W}$ *given by (4.35) provides unbiased estimates if and only if the matrix* $\tilde{\mathbf{\Theta}}^{AB}$ *is a standardised link matrix.*

Proof:

Starting from (4.35), we have

$$E(\mathbf{W}) = \tilde{\mathbf{\Theta}}^{T,AB}\mathbf{1}^A. \tag{4.36}$$

Using Theorem 4.3, we directly get $E(\mathbf{W}) = \mathbf{1}^B$ and therefore we have unbiased estimates. Now, assume that $E(\mathbf{W}) = \mathbf{1}^B$. From (4.36), we must have $\tilde{\mathbf{\Theta}}^{T,AB}\mathbf{1}^A = \mathbf{1}^B$ and therefore, $\tilde{\mathbf{\Theta}}^{AB}$ is a standardised link matrix. ■

The variance (4.11a) of $\hat{Y}^B$ is here expressed as

$$\begin{aligned} Var(\hat{Y}^B) &= \mathbf{Z}^T \mathbf{\Delta}^A \mathbf{Z} \\ &= \mathbf{Y}^{T,B} \mathbf{\Delta}^B \mathbf{Y} \end{aligned} \tag{4.37}$$

where $\mathbf{\Delta}^A = [(\pi_{jj'}^A - \pi_j^A \pi_{j'}^A) / \pi_j^A \pi_{j'}^A]_{N^A \times N^A}$ is a non-negative definite matrix of size $N^A \times N^A$ and where $\pi_{jj'}^A$ is the joint selection probability of units j and j' from U^A, and where $\Delta^B = \tilde{\mathbf{\Theta}}^{T,AB} \mathbf{\Delta}^A \tilde{\mathbf{\Theta}}^{AB}$.

As shown in Theorem 4.4, the estimator $\hat{Y}^B$ obtained by the GWSM will provide unbiased estimates provided that the matrix $\tilde{\mathbf{\Theta}}^{AB}$ is a standardised link matrix. Now, given that the variance (4.37) of this estimator depends on this matrix, there should be at least one matrix $\tilde{\mathbf{\Theta}}^{AB,opt}$ such that the variance of the estimator $\hat{Y}^B$ will be minimal. That is, for the $\theta_{j,i}^{AB}$ that are greater than 0, we are interested in finding the values that these $\theta_{j,i}^{AB}$ should have to obtain the most precise estimator $\hat{Y}^B$.

This optimality problem was first assessed by Kalton and Brick (1995). They obtained results based on the simplified situation where N^A=2 and with s^A obtained through equal probability sampling. Their conclusions suggested the use of $\theta_{ji}^{AB,opt} = 1$ when $\theta_{ji}^{AB} > 0$, and $\theta_{ji}^{AB,opt} = 0$ when $\theta_{ji}^{AB} = 0$. Lavallée (2002) and Lavallée and Caron (2001) obtained results along the same lines by the use of simulations.

In order to find the optimal weighted links $\theta_{j,i}^{AB,opt}$, we need to first factorise the standardised link matrix $\tilde{\mathbf{\Theta}}^{AB}$. *Factorisation* consists in finding a population U^G and standardised link matrices $\tilde{\mathbf{\Theta}}^{AG}$ and $\tilde{\mathbf{\Theta}}^{GB}$ such that $\tilde{\mathbf{\Theta}}^{AB} = \tilde{\mathbf{\Theta}}^{AG}\tilde{\mathbf{\Theta}}^{GB}$. We consider the population U^G containing as many units as there are links starting from the units j of U^A. The population size N^G is then given by the number of $\theta_{j,i}^{AB}$ from $\mathbf{\Theta}^{AB}$ that are greater than 0. Each unit g of U^G can be seen as the extremity of an "arrow" starting from some unit j of U^A. From this graph, there is only one link matrix $\mathbf{\Theta}^{AG}$ of size $N^A \times N^G$ keeping unbiasedness, namely $\mathbf{\Theta}^{AG} = [\theta_{j,g}^{AG}]$ where $\theta_{j,g}^{AG} = 1$ if there is a link (or an "arrow") leaving unit j of U^A to unit g from U^G, and $\theta_{j,g}^{AG} = 0$ otherwise. Note that by construction, each unit g from U^G is linked to at most one unit j from U^A and therefore $\tilde{\mathbf{\Theta}}^{AG} = \mathbf{\Theta}^{AG}$.

Considering the graph from U^G to U^B, we can construct the link matrix $\mathbf{\Theta}^{GB}$ of size $N^G \times N^B$ as follows. Because of the definition of the population U^G, each unit g of U^G is linked to exactly one unit i of U^B. Let $\tilde{\mathbf{\Theta}}^{GB} = \mathbf{\Theta}^{GB}[diag(\mathbf{1}^{T,G}\mathbf{\Theta}^{GB})]^{-1}$ be the standardised link matrix obtained from $\mathbf{\Theta}^{GB}$. We have $diag(\mathbf{1}^{T,G}\mathbf{\Theta}^{GB}) = diag(\mathbf{1}^{T,A}\mathbf{\Theta}^{AB})$, and therefor $\tilde{\mathbf{\Theta}}^{GB} = \mathbf{\Theta}^{GB}[diag.(\mathbf{1}^{T,A}\mathbf{\Theta}^{AB})]^{-1}$

Now,

$$
\begin{aligned}
\tilde{\mathbf{\Theta}}^{AG}\tilde{\mathbf{\Theta}}^{GB} &= \mathbf{\Theta}^{AG}\tilde{\mathbf{\Theta}}^{GB} \\
&= \mathbf{\Theta}^{AG}\mathbf{\Theta}^{GB}[diag(\mathbf{1}^{T,A}\mathbf{\Theta}^{AB})]^{-1} \\
&= \mathbf{\Theta}^{AB}[diag(\mathbf{1}^{T,A}\mathbf{\Theta}^{AB})]^{-1} \\
&= \tilde{\mathbf{\Theta}}^{AB}
\end{aligned}
\tag{4.38}
$$

Therefore, using this construction, the standardised link matrix $\widetilde{\mathbf{\Theta}}^{AB}$ from U^A to U^B can always be factorised into $\widetilde{\mathbf{\Theta}}^{AG}$ and $\widetilde{\mathbf{\Theta}}^{GB}$.

Using the factorisation (4.38), we have

$$\begin{aligned} Var(\hat{Y}^B) &= \mathbf{Y}^{T,B}\widetilde{\mathbf{\Theta}}^{T,AB}\mathbf{\Delta}^A\widetilde{\mathbf{\Theta}}^{AB}\mathbf{Y}^B \\ &= \mathbf{Y}^{T,B}\widetilde{\mathbf{\Theta}}^{T,GB}\widetilde{\mathbf{\Theta}}^{T,AG}\mathbf{\Delta}^A\widetilde{\mathbf{\Theta}}^{AG}\widetilde{\mathbf{\Theta}}^{GB}\mathbf{Y}^B \\ &= \mathbf{Y}^{T,B}\widetilde{\mathbf{\Theta}}^{T,GB}\mathbf{\Delta}^G\widetilde{\mathbf{\Theta}}^{GB}\mathbf{Y}^B \end{aligned} \tag{4.39}$$

where $\mathbf{\Delta}^G = \widetilde{\mathbf{\Theta}}^{T,AG}\mathbf{\Delta}^A\widetilde{\mathbf{\Theta}}^{AG}$.

For any standardised link matrix $\widetilde{\mathbf{\Theta}}^{AB}$, the factorisation (4.38) always produces the same first factor $\widetilde{\mathbf{\Theta}}^{AG}$. Therefore, if we seek some *optimal link matrix* $\widetilde{\mathbf{\Theta}}^{AB,opt}$ that minimises the variance (4.37), it is sufficient to optimise the second factor $\widetilde{\mathbf{\Theta}}^{GB}$. We would also like the optimal matrix $\widetilde{\mathbf{\Theta}}^{AB,opt}$ to produce unbiased estimates.

Let U_i^G be the subpopulation of U^G containing the N_i^G links to unit i of U^B. Note that the subpopulations U_i^G are disjoint. Thus, without loss of generality, we can order the links from U^A to U^B so that, for every i, the links to unit i in U^B are indexed consecutively. Now, let $\dot{\widetilde{\mathbf{\theta}}}_i^{GB}$ be the ith column vector of the matrix $\widetilde{\mathbf{\Theta}}^{GB}$, i=1,…,N^B. By construction, the vector $\dot{\widetilde{\mathbf{\theta}}}_i^{GB}$ contains non null elements only for the N_i^G links to unit i of U^B. Hence, letting $\widetilde{\mathbf{\theta}}_i^{GB}$ be a column vector of size N_i^G containing the non null elements of $\dot{\widetilde{\mathbf{\theta}}}_i^{GB}$, we have $\dot{\widetilde{\mathbf{\theta}}}_i^{GB} = \begin{bmatrix} \mathbf{0} \\ \widetilde{\mathbf{\theta}}_i^{GB} \\ \mathbf{0} \end{bmatrix}$. Similarly, let $\dot{\mathbf{1}}_i^G$ be the column vector of size N^G containing 1's for N_i^G elements and 0's elsewhere. Letting $\mathbf{1}_i^G$ be a column vector of size N_i^G containing 1's, we have $\dot{\mathbf{1}}_i^G = \begin{bmatrix} \mathbf{0} \\ \mathbf{1}_i^G \\ \mathbf{0} \end{bmatrix}$. Now, for the GWSM from U^G to U^B to be unbiased, we need to have

$\tilde{\boldsymbol{\theta}}_i^{T,GB}\mathbf{1}_i^G = 1$ for all *i*, or equivalently $\dot{\tilde{\boldsymbol{\theta}}}_i^{T,GB}\dot{\mathbf{1}}_i^G = 1$. All this together leads to the following optimisation problem:

FIND A MATRIX $\tilde{\boldsymbol{\Theta}}^{GB,opt} = \{\dot{\tilde{\boldsymbol{\theta}}}_1^{GB,opt},...,\dot{\tilde{\boldsymbol{\theta}}}_{N^B}^{GB,opt}\}$ SATISFYING $\dot{\tilde{\boldsymbol{\theta}}}_i^{T,GB}\dot{\mathbf{1}}_i^G = 1$ FOR ALL $i=1,\ldots,N^B$, AND MINIMISING THE QUADRATIC FORM $Var(\hat{Y}^B) = \mathbf{Y}^{T,B}\tilde{\boldsymbol{\Theta}}^{T,GB}\boldsymbol{\Delta}^G\tilde{\boldsymbol{\Theta}}^{GB}\mathbf{Y}^B$.

This problem turns out to be nothing more than the minimisation of a positive quadratic form under linear constraints. This is a relatively standard and simple problem to solve. It is well known that a solution always exists and is unique if the form (4.39) is positive definite or if the null subspace of $\tilde{\boldsymbol{\Theta}}^{GB}$ is not included in the null-space of $\boldsymbol{\Delta}^G$.

The above optimisation problem can be rewritten in a different form. Let $\boldsymbol{\Delta}_{ii'}^G$ be the submatrix of $\boldsymbol{\Delta}^G$ corresponding to the elements in positions g and g' if g has a link with unit i and g' has a link with unit i'. These matrices form a partition of $\boldsymbol{\Delta}^G$. Note that the matrices $\boldsymbol{\Delta}_{ii}^G$ are symmetric, positive definite, and $\boldsymbol{\Delta}_{ii'}^{T,G} = \boldsymbol{\Delta}_{i'i}^G$. With these notations, the optimisation problem can be written as:

MINIMISE

$$\sum_{i=1}^{N^B}\sum_{i'=1}^{N^B} y_i y_{i'} \tilde{\boldsymbol{\theta}}_i^{T,GB}\boldsymbol{\Delta}_{ii'}^G\tilde{\boldsymbol{\theta}}_{i'}^{GB} \tag{4.40}$$

UNDER THE CONSTRAINTS $\tilde{\boldsymbol{\theta}}_i^{T,GB}\mathbf{1}_i^G = 1$ FOR ALL $i=1,\ldots,N^B$.

Minimisation is achieved for vectors $\tilde{\boldsymbol{\theta}}_i^{GB,opt}$ satisfying

$$y_i\sum_{i'=1}^{N^B}\boldsymbol{\Delta}_{ii'}^G\tilde{\boldsymbol{\theta}}_{i'}^{GB,opt}y_{i'} = \lambda_i\mathbf{1}_i^G \tag{4.41}$$

for all $i=1,\ldots,N^B$ and where λ_i are the Lagrange multipliers entering into the constrained minimisation of (4.40). As we can see from (4.41), the optimal choice $\tilde{\boldsymbol{\theta}}_i^{GB,opt}$ (and therefore $\tilde{\boldsymbol{\Theta}}^{GB,opt}$) will depend, in general, explicitly on the vector $\mathbf{Y}^B$, which is not useful in practice. Notice that the set of λ_i depends also on the variable $\mathbf{Y}^B$. This is the reason why we should seek, instead of a strong optimisation, a weaker form of optimality that will lead to the existence of an "optimal" solution $\widetilde{\mathbf{\Theta}}^{GB,opt}$ (and $\widetilde{\mathbf{\Theta}}^{AB,opt}$) not depending on $\mathbf{Y}^B$.

Equations (4.41) must be valid for any vector $\mathbf{Y}^B$. A necessary condition is to have them hold a particular variable of interest, such as $y_i = 1$ for a unit i of U^B and $y_{i'} = 0$ for all other units i' of U^B ($i' \neq i$). This leads to the necessary conditions (one for each of those particular variables) $\mathbf{\Delta}_{ii}^G \widetilde{\mathbf{\theta}}_i^{GB,opt} = \lambda_i \mathbf{1}_i^G$. Assuming that $\mathbf{\Delta}_{ii}^G$ is invertible, we then have $\widetilde{\mathbf{\theta}}_i^{GB,opt} = \lambda_i (\mathbf{\Delta}_{ii}^G)^{-1} \mathbf{1}_i^G$. It can be shown that this is also a sufficient condition. Now, because $\widetilde{\mathbf{\theta}}_i^{T,GB,opt} \mathbf{1}_i^G = 1$, we have $\lambda_i = 1 / \mathbf{1}_i^{T,G} (\mathbf{\Delta}_{ii}^G)^{-1} \mathbf{1}_i^G$. Therefore, a necessary and sufficient condition for equation (4.41) to be satisfied is when

$$\widetilde{\mathbf{\theta}}_i^{GB,opt} = \frac{(\mathbf{\Delta}_{ii}^G)^{-1} \mathbf{1}_i^G}{\mathbf{1}_i^{T,G} \mathbf{\Delta}_{G,ii}^{-1} \mathbf{1}_i^G}. \tag{4.42}$$

This result corresponds to *weak optimality* as it holds for a particular variable of interest.

Weak optimality is a necessary condition for *strong optimality* independent of the vector $\mathbf{Y}^B$ for a variable of interest. It provides the necessary form for the vectors $\widetilde{\mathbf{\theta}}_i^{GB,opt}$ in (4.41). To get sufficient conditions for strong optimality independent of $\mathbf{Y}^B$, we go back to equations (4.41). These equations need to be satisfied for all vectors $\mathbf{Y}^B$ and they must therefore be satisfied for a particular variable of interest such as $y_i = 1$ for a unit i of U^B, $y_{i'} = 1$ for another unit i' of U^B, and $y_{i''} = 0$ for all other units i'' of U^B ($i'' \neq i' \neq i$). In that case, to satisfy equations (4.41), it is necessary to have the following relations for any i and i':

$$\mathbf{\Delta}_{ii}^{G}\tilde{\mathbf{\theta}}_{i}^{GB,opt} + \mathbf{\Delta}_{ii'}^{G}\tilde{\mathbf{\theta}}_{i'}^{GB,opt} = \lambda_{i}^{ii'}\mathbf{1}_{i}^{G} \tag{4.43}$$

$$\mathbf{\Delta}_{i'i'}^{G}\tilde{\mathbf{\theta}}_{i'}^{GB,opt} + \mathbf{\Delta}_{i'i}^{G}\tilde{\mathbf{\theta}}_{i}^{GB,opt} = \lambda_{i'}^{ii'}\mathbf{1}_{i'}^{G} .$$

As we must necessarily have weak optimality, we have $\mathbf{\Delta}_{ii}^{G}\tilde{\mathbf{\theta}}_{i}^{GB,opt} = \lambda_{i}\mathbf{1}_{i}^{G}$. Considering the first line of (4.43), we then get

$$\begin{aligned}\mathbf{\Delta}_{ii'}^{G}\tilde{\mathbf{\theta}}_{i'}^{GB,opt} &= (\lambda_{i}^{ii'} - \lambda_{i})\mathbf{1}_{i}^{G} \\ &= \Phi_{ii'}\mathbf{1}_{i}^{G}\end{aligned} \tag{4.44}$$

Multiplying both sides of (4.44) by $\tilde{\mathbf{\theta}}_{i}^{T,GB,opt}$, we obtain

$$\begin{aligned}\tilde{\mathbf{\theta}}_{i}^{T,GB,opt}\mathbf{\Delta}_{ii'}^{G}\tilde{\mathbf{\theta}}_{i'}^{GB,opt} &= \Phi_{ii'}\tilde{\mathbf{\theta}}_{i}^{T,GB,opt}\mathbf{1}_{i}^{G} \\ &= \Phi_{ii'}\end{aligned}$$

since $\tilde{\mathbf{\theta}}_{i}^{T,GB,opt}\mathbf{1}_{i}^{G} = 1$. Let $\mathbf{\Phi}$ be the matrix with elements $\Phi_{ii'}$ off the diagonal and $\Phi_{ii} = \lambda_{i}$ on the diagonal. Using again (4.39), it can be shown that the optimal variance (whenever it exists) has the expression $\mathbf{Y}^{T,B}\mathbf{\Phi}\mathbf{Y}^{B}$.

Let us show that this set of conditions is also sufficient. Assume that (4.44) holds. Note that for $i = i'$, condition (4.44) is nothing more than (4.42) which gives the necessary values for the $\tilde{\mathbf{\theta}}_{i}^{GB,opt}$. It is now straightforward to verify that (4.41) holds, whatever the value of $\mathbf{Y}^{B}$, and that we have obtained strong optimality. Now, the values of λ_{i} depend on $\mathbf{Y}^{B}$, as well as the variance $Var(\hat{Y}^{B})$, but we have that equations (4.41) always have the same solution (4.42) that does not depend on $\mathbf{Y}^{B}$. We therefore have the following theorem:

Theorem 4.5: Strong optimisation independent of $\mathbf{Y}^{B}$

The conditions $\mathbf{\Delta}_{ii'}^{G}\tilde{\mathbf{\theta}}_{i'}^{GB,opt} = \Phi_{ii'}\mathbf{1}_{i}^{G}$ *are necessary and sufficient for the existence of a standardised link matrix* $\tilde{\mathbf{\Theta}}^{GB,opt}$, *or equivalently* $\tilde{\mathbf{\Theta}}^{AB,opt}$, *that achieves strong optimality independent of the vector* $\mathbf{Y}^{B}$ *for the variable of interest. The values in the columns of this strong optimal matrix are given by (4.42), which are the vectors* $\tilde{\mathbf{\theta}}_{i}^{GB,opt}$ *obtained from weak optimality.*

Since $\mathbf{\Delta}_{ii}^{G}\widetilde{\mathbf{\theta}}_{i}^{GB,opt} = \lambda_i \mathbf{1}_i^G$, (4.44) can be written equivalently as

$$\Phi_{ii'}^{**}\ \widetilde{\mathbf{\theta}}_{i}^{GB,opt} = (\mathbf{\Delta}_{ii}^{G})^{-1}\mathbf{\Delta}_{ii'}^{G}\widetilde{\mathbf{\theta}}_{i'}^{GB,opt} \tag{4.45a}$$

or

$$\Phi_{ii'}^{*}\ \mathbf{1}_i^G = \mathbf{\Delta}_{ii'}^{G}(\mathbf{\Delta}_{i'i'}^{G})^{-1}\mathbf{1}_{i'}^G \tag{4.45b}$$

where $\Phi_{ii'}^{**} = (\widetilde{\mathbf{\theta}}_{i}^{T,GB,opt}\ \mathbf{\Delta}_{ii'}^{G}\widetilde{\mathbf{\theta}}_{i'}^{GB,opt})\ (\mathbf{1}_i^{T,G}(\mathbf{\Delta}_{ii}^{G})^{-1}\ \mathbf{1}_i^G)$ and $\Phi_{ii'}^{*}$ = $(\widetilde{\mathbf{\theta}}_{i}^{T,GB,opt}\ \mathbf{\Delta}_{ii'}^{G}\widetilde{\mathbf{\theta}}_{i'}^{GB,opt})\ (\mathbf{1}_{i'}^{T,G}(\mathbf{\Delta}_{i'i'}^{G})^{-1}\ \mathbf{1}_{i'}^G)$. In some situations, these can proved to be easier to use than the expresssion (6.7) stated in Theorem 4.5.

We now present an example that illustrates the preceding theory on weak optimality and strong optimality independent of $\mathbf{Y}^B$.

***Example* 4.3**

Let us suppose that the sample s^A is selected using simple random sampling. In that case, the $N^A \times N^A$ matrix $\mathbf{\Delta}^A$ is given by $\mathbf{\Delta}^A = \frac{N^A}{n^A}\frac{(N^A - n^A)}{(N^A - 1)}\left[\mathbf{I}^A - \frac{\mathbf{1}^A\mathbf{1}^{T,A}}{N^A}\right]$ where $\mathbf{I}^A$ is the identity matrix of size $N^A \times N^A$. Considering the factorisation (4.38), we have

$$\begin{aligned}\mathbf{\Delta}^G &= \widetilde{\mathbf{\Theta}}^{T,AG}\mathbf{\Delta}^A\widetilde{\mathbf{\Theta}}^{AG}\\ &= \frac{N^A}{n^A}\frac{(N^A - n^A)}{(N^A - 1)}\times\widetilde{\mathbf{\Theta}}^{T,AG}\left[\mathbf{I}^A - \frac{\mathbf{1}^A\mathbf{1}^{T,A}}{N^A}\right]\widetilde{\mathbf{\Theta}}^{AG}\\ &= \frac{N^A}{n^A}\frac{(N^A - n^A)}{(N^A - 1)}\times\left[diag(\mathbf{1}_{jj}^A) - \frac{\mathbf{1}^G\mathbf{1}^{T,G}}{N^A}\right]\end{aligned} \qquad 4.46)$$

where $\mathbf{1}_{jj}^A$ is a square matrix of size N_j^A, with N_j^A being the number of links (or "arrows") starting from unit j of U^A. From $\mathbf{\Delta}^G$, we extract the submatrices $\mathbf{\Delta}_{ii}^G$. Each submatrix $\mathbf{\Delta}_{ii}^G$ is given by $\mathbf{\Delta}_{ii}^G = \frac{N^A}{n^A}\frac{(N^A - n^A)}{(N^A - 1)}\times\left[\mathbf{I}_i^G - \frac{\mathbf{1}_i^G\mathbf{1}_i^{T,G}}{N^A}\right]$, which is of size N_i^G. Then, using a matrix result that can be found in Jazwinski (1970), we get

$(\boldsymbol{\Delta}_{ii}^{G})^{-1} = \frac{(N^A-1)}{(N^A-n^A)}\frac{n^A}{N^A} \times \left[\mathbf{I}_i^G + \frac{1}{(N^A-N_i^G)} \mathbf{1}_i^G \mathbf{1}_i^{T,G} \right]$. Now, from (4.42), we directly obtain the optimal values $\tilde{\boldsymbol{\theta}}_i^{GB,opt} = \frac{1}{N_i^G}\mathbf{1}_i^G$ that minimise $Var(\hat{Y}^B)$, in the weak sense, $i = 1,...,N^B$. These values are used to construct the vectors $\dot{\tilde{\boldsymbol{\theta}}}_i^{T,GB,opt}$, and then the matrix $\tilde{\boldsymbol{\Theta}}_{GB,opt} = \{\dot{\tilde{\boldsymbol{\theta}}}_1^{GB,opt},...,\dot{\tilde{\boldsymbol{\theta}}}_{N^B}^{GB,opt}\}$. Finally, after computing the optimal matrix $\tilde{\boldsymbol{\Theta}}^{AB,opt} = \boldsymbol{\Theta}^{AG}\tilde{\boldsymbol{\Theta}}^{GB,opt}$, we obtain the optimal weights $\mathbf{W}^{opt}$ using (4.35).

Again, this result is an important one because it goes directly in the direction of the results of Kalton and Brick (1995), Lavallée (2002), and Lavallée and Caron (2001). That is, with simple random sampling, the optimal choice of $\theta_{j,i}^{AB,opt}$ should be 1 if there is a link between unit j of U^A and i of U^B, and 0 otherwise.

Using Theorem 4.5, we now verify if the conditions (4.44), (4.45a) or (4.45b) for strong optimality independent of y_i are satisfied for the optimal matrix $\tilde{\boldsymbol{\Theta}}^{AB,opt}$ that we obtain through weak optimisation. First, each submatrix $\Delta_{ii'}^G$ of size $N_i^G \times N_{i'}^G$ is given by $\boldsymbol{\Delta}_{ii'}^G = \frac{N^A}{n^A}\frac{(N^A-n^A)}{(N^A-1)} \times \left[\mathbf{H}_{ii'}^G - \frac{\mathbf{1}_i^G\mathbf{1}_{i'}^{T,G}}{N^A} \right]$ where $\mathbf{H}_{ii'}^G$ is a $N_i^G \times N_{i'}^G$ diagonal matrix of ones, "padded" with zeros. A typical element of $\mathbf{H}_{ii'}^G$ is given by 1 if both i and i' are linked to the same unit j of U^A (that is linked to unit g of U^G), and 0 otherwise. Therefore, we can easily see in which cases the conditions (4.44), (4.45a) or (4.45b) can be satisfied. In fact, because all components of $\tilde{\boldsymbol{\theta}}_i^{GB,opt}$ are equal, $\boldsymbol{\Delta}_{ii'}^G\tilde{\boldsymbol{\theta}}_{i'}^{GB,opt}$ is a vector proportional to the sum of the lines of $\boldsymbol{\Delta}_{ii'}^G$, i.e., the sum of the lines of $\left[\mathbf{H}_{ii'}^G - \frac{\mathbf{1}_i^G\mathbf{1}_{i'}^{T,G}}{N^A} \right]$. But (4.44) says that this vector must have the same components. This is possible if and only if the matrix $\mathbf{H}_{ii'}^G$ contains only zeros, or if it is of dimension 1×1, which occurs when both i and i' are each linked to only one element of U^A. Therefore, strong optimality independent of $\mathbf{Y}^B$ does not occur in general for simple random sampling.

CHAPTER 5

OTHER GENERALISATIONS

We mentioned in Chapter 1 that the GWSM is in fact a generalisation of the weight share method described by Ernst (1989). It can also be considered as a generalisation of network sampling as well as adaptive cluster sampling described by Thompson (1992), Thompson and Seber (1996) and Thompson (2002). It is however possible to go a little further with the by expanding the context in which the GWSM can be used. Firstly, we will consider the possibility of performing a two-stage indirect sampling. Secondly, we will discuss the arbitrary aspect of the formation of clusters. Finally, we will examine the possibility of eliminating the notion of clusters.

5.1 TWO-STAGE INDIRECT SAMPLING

An important constraint to which the survey process was subjected is to consider all units belonging to the same cluster. In other words, if a unit is selected in the sample, then all units from the cluster containing the selected unit must be surveyed. Although this constraint often permits savings and also allows for estimates on the clusters to be produced, we may want to consider only a subsample of units from the cluster to survey. This could turn out to be useful, for example, when the cluster size is considerable.

As an example, we can consider the survey of enterprises through their establishments, as shown in Figure 1.3. Recall that we select a sample of establishments, we go to the enterprise level (cluster) and we finally survey all establishments from the identified enterprises. Unfortunately, the number of establishments for certain enterprises (like chains of small retail stores, for example) can prove to be enormous and in this case, we may want to restrict ourselves to a subsample of establishments. We could however argue that we already had a sample of establishments at the beginning, and why then would we select another one? In the example that concerns us, this new sample allows us to give a non-zero selection probability to establishments **f** and **g** that have no chance of being selected in the sample at the beginning.

In a formal way, a sample s^A is selected as before containing m^A units from the population U^A containing M^A units according to a certain sampling design. We assume that π_j^A represents the selection probability of unit j and that $\pi_j^A > 0$ for all $j \in U^A$. On the other hand, the target population U^B contains M^B units. This population is divided into N clusters, where cluster i contains M_i^B units.

For each unit j selected in s^A, we identify the units ik from U^B that have a non-zero link with j, i.e., $l_{j,ik} = 1$. For each identified unit ik, we assume that we can create the list of M_i^B units of cluster i containing this unit. This cluster i itself then represents a population U_i^B where $U^B = \bigcup_{i=1}^N U_i^B$. Let Ω^B be the set of n clusters identified by the units $j \in s^A$.

From each cluster $i \in \Omega^B$, a sample s_i^B containing m_i^B units is selected from the M_i^B units of the cluster. We assume that $\pi_{(i)k}^{II}$ represents the selection probability of unit k and $\pi_{(i)k}^{II} > 0$ for all $k \in U_i^B$. The variable of interest y is measured only for the units from the samples s_i^B, $i=1,...n$.

In the context of indirect sampling presented in Chapter 2, we had performed a census of each cluster $i \in \Omega^B$. The fact of selecting a sample s_i^B of clusters i from Ω^B brings us a second stage to the sampling design. Since the first stage of the sampling design is an indirect sampling, we can call the present design *two-stage indirect sampling*. Note that a similar two-stage design was proposed by Sirken and Shimizu (1999) in the context of network sampling.

By applying the GWSM, we want to assign an estimation weight w_{ik}^{II} to each unit $k \in s_i^B$ of the n clusters $i \in \Omega^B$. To estimate the total Y^B of the target population U^B, we can then use the estimator

$$\hat{Y}^{II,B} = \sum_{i=1}^{n}\sum_{k=1}^{m_i^B} w_{ik}^{II} y_{ik} . \quad (5.1)$$

We now present the steps of the GWSM to obtain the weights w_{ik}^{II}.

Steps of the GWSM for two-stage indirect sampling

Step 1: For each unit k of the M_i^B units from cluster i contributing to $\hat{Y}^{II,B}$, calculate the initial weight w'_{ik}, that is:

$$w'_{ik} = \sum_{j=1}^{M^A} l_{j,ik} \frac{t_j}{\pi_j^A}, \quad (5.2)$$

where $t_j = 1$ if $j \in s^A$, and 0 otherwise.

Step 2: For each unit k of the M_i^B units from cluster i contributing to $\hat{Y}^{II,B}$, obtain the total number of links L_{ik}^B:

$$L_{ik}^B = \sum_{j=1}^{M^A} l_{j,ik} . \quad (5.3)$$

Step 3: Calculate the first-stage weight w_i given by

$$w_{ik} = \frac{\sum_{k=1}^{M_i^B} w'_{ik}}{\sum_{k=1}^{M_i^B} L_{ik}^B} . \quad (5.4)$$

Step 4: Finally, set $w_{ik}^{II} = w_i / \pi_{(i)k}^{II}$ for all $k \in s_i^B$.

In order to calculate the bias and the variance of $\hat{Y}^{II,B}$, we prove the following Theorem 5.1, inspired by Theorem 4.1.

Theorem 5.1: Duality in the form of $\hat{Y}^{II,B}$

*Let $\hat{Y}_i = \sum_{k=1}^{m_i^B} y_{ik} / \pi_{(i)k}^{II}$ and $L_i^B = \sum_{k=1}^{M_i^B} L_{ik}^B$. For the clusters $i \in \Omega^B$, set $\hat{z}_{ik} = \hat{Y}_i / L_i^B$ for **all** units $k \in U_i^B$. The estimator $\hat{Y}^{II,B}$, given by (5.1), can then also be written in the form*

$$\hat{Y}^{II,B} = \sum_{j=1}^{M^A} \frac{t_j}{\pi_j^A} \hat{Z}_j , \quad (5.5)$$

where
$$\hat{Z}_j = \sum_{i=1}^{n}\sum_{k=1}^{M_i^B} l_{j,ik}\hat{z}_{ik} \; . \tag{5.6}$$

Proof

From $\hat{Y}^{II,B} = \sum_{i=1}^{n} w_i \sum_{k=1}^{m_i^B} y_{ik} / \pi_{(i)k}^{II} = \sum_{i=1}^{n} w_i \hat{Y}_i$, we substitute the definition of w_i in $\hat{Y}^{II,B}$ to get:

$$\hat{Y}^{II,B} = \sum_{i=1}^{n} \hat{Y}_i \left(\frac{\sum_{k=1}^{M_i^B} w'_{ik}}{\sum_{k=1}^{M_i^B} L_{ik}^B} \right) = \sum_{i=1}^{n} \frac{\hat{Y}_i}{L_i^B} \sum_{k=1}^{M_i^B} w'_{ik} \; . \tag{5.7}$$

Let $\hat{z}_{ik} = \hat{Y}_i / L_i^B$. Note that this quantity is defined if and only if $L_i^B > 0$, that is, if and only if Constraint 2.1 is satisfied. We then get

$$\hat{Y}^B = \sum_{i=1}^{n}\sum_{k=1}^{M_i^B} w'_{ik}\hat{z}_{ik} \; . \tag{5.8}$$

By replacing w'_{ik} with its definition (5.2), we get

$$\begin{aligned} \hat{Y}^B &= \sum_{i=1}^{n}\sum_{k=1}^{M_i^B} \left(\sum_{j=1}^{M^A} l_{j,ik} \frac{t_j}{\pi_j^A} \right) \hat{z}_{ik} \\ &= \sum_{i=1}^{n}\sum_{k=1}^{M_i^B}\sum_{j=1}^{M^A} l_{j,ik} t_j \frac{\hat{z}_{ik}}{\pi_j^A}. \end{aligned} \tag{5.9}$$

Finally,

$$\begin{aligned} \hat{Y}^B &= \sum_{j=1}^{M^A} \frac{t_j}{\pi_j^A} \sum_{i=1}^{n}\sum_{k=1}^{M^B} l_{j,ik}\hat{z}_{ik} \\ &= \sum_{j=1}^{M^A} \frac{t_j}{\pi_j^A} \hat{Z}_j . \end{aligned} \tag{5.10}$$

■

The estimator $\hat{Y}^{II,B}$ can therefore be written as a function of the units *ik* of U^B, or as a function of the units *j* of U^A.

Corollary 5.1: Bias of $\hat{Y}^{II,B}$

The estimator $\hat{Y}^{II,B}$ is unbiased for the estimation of Y^B, with respect to the sampling design.

Proof

Take the expectation $E(\hat{Y}^{II,B})$ from (5.5), with respect to the design. This expectation can be decomposed into $E_{\Omega^B}[E(\hat{Y}^{II,B} \mid \Omega^B)]$, where the first expectation is performed with respect to all possible samples Ω^B of clusters, and the second expectation is conditional on the clusters of Ω^B. From (5.5) and (5.6), we have

$$E(\hat{Y}^{II,B} \mid \Omega^B) = \sum_{j=1}^{M^A} \frac{t_j}{\pi_j^A} \sum_{i=1}^{n} \sum_{k=1}^{M^B} l_{j,ik} E(\hat{z}_{ik} \mid \Omega^B). \tag{5.11}$$

Now,

$$\begin{aligned} E(\hat{z}_{ik} \mid \Omega^B) &= E\left(\frac{\hat{Y}_i}{L_i^B} \mid \Omega^B\right) = \frac{1}{L_i^B} E\left(\sum_{k=1}^{m_i^B} \frac{y_{ik}}{\pi_{(i)k}^{II}} \mid \Omega^B\right) \\ &= \frac{Y_i}{L_i^B} = z_{ik} \end{aligned} \tag{5.12}$$

since $\hat{Y}_i = \sum_{k=1}^{m_i^B} y_{ik} / \pi_{ik}^B$ is nothing more than a Horvitz-Thompson estimator of Y_i. Thus,

$$E(\hat{Y}^{II,B} \mid \Omega^B) = \sum_{j=1}^{M^A} \frac{t_j}{\pi_j^A} \sum_{i=1}^{n} \sum_{k=1}^{M^B} l_{j,ik} z_{ik}. \tag{5.13}$$

Following the survey process, the cluster i will be part of the n clusters of Ω^B if and only if $l_{j,ik} \neq 0$ (there is a link between units j of U^A and ik of U^B) for at least one $k \in U_i^B$, and $t_j = 1$ (unit j of U^A is selected in s^A), or in other words, if and only if $l_{j,ik} t_j \neq 0$. Unit k of cluster i is therefore surveyed if and only if, for all π_j^A, we have $l_{j,ik} t_j / \pi_j^A \neq 0$ for at least one $k \in U_i^B$, which implies that $\varphi_i = \sum_{k=1}^{M_i^B} \sum_{j=1}^{M^A} l_{j,ik} t_j z_{ik} / \pi_j^A \neq 0$. The n clusters surveyed therefore have $\varphi_i \neq 0$, and the N-n non-surveyed clusters have $\varphi_i = 0$. Thus,

$$\begin{aligned} E(\hat{Y}^{II,B} \mid \Omega^B) &= \sum_{j=1}^{M^A} \frac{t_j}{\pi_j^A} \sum_{i=1}^{n} \sum_{k=1}^{M^B} l_{j,ik} z_{ik} \\ &= \sum_{j=1}^{M^A} \frac{t_j}{\pi_j^A} \sum_{i=1}^{N} \sum_{k=1}^{M^B} l_{j,ik} z_{ik} = \hat{Y}^B. \end{aligned} \tag{5.14}$$

From Corollary 4.2, we have directly $E_{\Omega^B}(\hat{Y}^B)=Y^B$ and therefore

$$E(\hat{Y}^{II,B})=Y^B. \tag{5.15}$$

■

Corollary 5.2: Variance of $\hat{Y}^{II,B}$

The variance formula of the estimator $\hat{Y}^{II,B}$*, with respect to the sampling design, is given by*

$$Var(\hat{Y}^{II,B})=\sum_{j=1}^{M^A}\sum_{i=1}^{N}\left(\frac{L_{j,i}}{L_i^B}\right)^2\sigma_i^2+\sum_{j=1}^{M^A}\sum_{j'=1}^{M^A}\frac{(\pi_{jj'}^A-\pi_j^A\pi_{j'}^A)}{\pi_j^A\pi_{j'}^A}Z_jZ_{j'}, \tag{5.16}$$

where

$$\sigma_i^2=\sum_{k=1}^{M_i^B}\sum_{k'=1}^{M_i^B}\frac{(\pi_{(i)k,(i)k'}^{II}-\pi_{(i)k}^{II}\pi_{(i)k'}^{II})}{\pi_{(i)k}^{II}\pi_{(i)k'}^{II}}y_{ik}y_{ik'} \tag{5.17}$$

and where $\pi_{(i)k,(i)k'}^{II}$ *represents the joint selection probability of units* k *and* k' *from cluster i.*

Proof

To get a variance formula for $\hat{Y}^{II,B}$, we start from equation (5.5). As with Corollary 5.1, we start from a conditional argument using the following identity from Särndal, Swensson and Wretman (1992):

$$Var(\hat{Y}^{II,B})=E_{\Omega^B}[Var(\hat{Y}^{II,B}\mid\Omega^B)]+Var_{\Omega^B}[E(\hat{Y}^{II,B}\mid\Omega^B)].$$

From equation (5.14) and from Corollary 4.2, we get directly

$$Var_{\Omega^B}[E(\hat{Y}^{II,B}\mid\Omega^B)]=\sum_{j=1}^{M^A}\sum_{j'=1}^{M^A}\frac{(\pi_{jj'}^A-\pi_j^A\pi_{j'}^A)}{\pi_j^A\pi_{j'}^A}Z_jZ_{j'}. \tag{5.18}$$

Now, from (5.5) and (5.6), we have

$$\begin{aligned}\hat{Y}^{II,B}&=\sum_{j=1}^{M^A}\frac{t_j}{\pi_j^A}\sum_{i=1}^{n}\sum_{k=1}^{M^B}l_{j,ik}\frac{\hat{Y}_i}{L_i^B}\\&=\sum_{j=1}^{M^A}\frac{t_j}{\pi_j^A}\sum_{i=1}^{n}\frac{\hat{Y}_i}{L_i^B}L_{j,i}.\end{aligned} \tag{5.19}$$

We then calculate the conditional variance of $\hat{Y}^{II,B}$ to get

$$Var(\hat{Y}^{II,B} \mid \Omega^B) = \sum_{j=1}^{M^A} \frac{t_j}{\pi_j^A} \sum_{i=1}^{n} \left(\frac{L_{j,i}}{L_i^B} \right)^2 Var(\hat{Y}_i \mid \Omega^B). \tag{5.20}$$

Since $\hat{Y}_i = \sum_{k=1}^{m_i^B} y_{ik} / \pi_{(i)k}^{II}$ is nothing more than a Horvitz-Thompson estimator of Y_i, we have

$$\begin{aligned} Var(\hat{Y}_i \mid \Omega^B) &= \sum_{k=1}^{M_i^B} \sum_{k'=1}^{M_i^B} \frac{\left(\pi_{(i)k,(i)k'}^{II} - \pi_{(i)k}^{II} \pi_{(i)k'}^{II} \right)}{\pi_{(i)k}^{II} \pi_{(i)k'}^{II}} y_{ik} y_{ik'} \\ &= \sigma_i^2 . \end{aligned} \tag{5.21}$$

From (5.20) and (5.21), using the same arguments as those used in obtaining (5.14), we have

$$Var(\hat{Y}^{II,B} \mid \Omega^B) = \sum_{j=1}^{M^A} \frac{t_j}{\pi_j^A} \sum_{i=1}^{n} \left(\frac{L_{j,i}}{L_i^B} \right)^2 \sigma_i^2 = \sum_{j=1}^{M^A} \frac{t_j}{\pi_j^A} \sum_{i=1}^{N} \left(\frac{L_{j,i}}{L_i^B} \right)^2 \sigma_i^2 . \tag{5.22}$$

Finally,

$$E_{\Omega^B} \left[Var(\hat{Y}^{II,B} \mid \Omega^B) \right] = \sum_{j=1}^{M^A} \sum_{i=1}^{N} \left(\frac{L_{j,i}}{L_i^B} \right)^2 \sigma_i^2 . \tag{5.23}$$

■

5.2 ARBITRARY ASPECT IN THE FORMATION OF CLUSTERS

We saw that the GWSM relates to an indirect sampling of clusters surveyed within the population U^B. In practice, the clusters of the population U^B are, most of the time, formed in a natural manner. In social surveys, for example, clusters often correspond to households or families, and the units are the people belonging to these households or these families. For economic surveys, the clusters often represent enterprises while the units of these clusters are establishments or local units. To apply the GWSM, the formation of the clusters can however be performed in an arbitrary fashion.

If the process of forming the clusters is independent from the selection of the sample s^A, the GWSM remains unbiased for the estimation of the total Y^B. We indeed notice that the proof of Corollary 4.1 does not mention the construction of the clusters themselves. Note, however, that for the GWSM to remain unbiased, the process of forming the clusters must respect Constraint 2.1. Assuming that the clusters respect Constraint 2.1, the choice of the clusters, however, will influence the precision of the estimates produced for the target population U^B. In other words, the variance of the estimator $\hat{Y}^B$ depends on the formation of the clusters.

In the construction of the clusters for the population U^B, we find two extreme cases: (i) the formation of a single cluster of size M^B and (ii) the formation of M^B clusters of size 1. Of course, in practice, the formation of the clusters is somewhere between these two extremes. Meanwhile, these two extreme cases can help us to understand the process governing the precision of the estimates according to the construction of the clusters.

5.2.1 Extreme case (i): population U^B with a single cluster of size M^B

Suppose that we decide to create a single cluster of size M^B for the target population U^B (Figure 5.1). Since the survey process requires us to survey all units belonging to the clusters selected indirectly through the sample s^A, we will then inevitably have a census of the population U^B. Indeed, for each unit j selected in s^A, we are linked to the population U^B by the links $l_{j,ik} > 0$. For each of these links, we survey all units of each linked cluster i and, as the population only has one single cluster, we will thus survey the entire population U^B.

As we are interested in the variance of the estimator $\hat{Y}^B$, it is practical to use the form of $\hat{Y}^B$ given by Corollary 4.3. Since U^B only has a single cluster i=1, we have $Y_1 = Y^B$, $L_{j,1} = \sum_{k=1}^{M_1^B} l_{j,1k} = L_j^A$, and $L_1^B = \sum_{k=1}^{M_1^B} L_{1k}^B = L$.

We then get

$$\begin{aligned} \hat{Y}^B &= \sum_{j=1}^{M^A} \frac{t_j}{\pi_j^A} \sum_{i=1} Y_i \frac{L_{j,i}}{L_i^B} = \sum_{j=1}^{M^A} \frac{t_j}{\pi_j^A} \frac{L_j^A}{L} Y^B \\ &= \frac{Y^B}{L} \sum_{j=1}^{M^A} \frac{t_j}{\pi_j^A} L_j^A = \frac{Y^B}{L} \hat{L} \end{aligned} \tag{5.24}$$

where $\hat{L}=\sum_{j=1}^{M^A} t_j L_j^A / \pi_j^A$. We can see that $\hat{L}$ is in fact a Horvitz-Thompson estimator of the total number of links L. The variance of $\hat{Y}^B$ in the case where the population U^B only has a single cluster is thus given by

$$Var(\hat{Y}^B)=\left(\frac{Y^B}{L}\right)^2 \sum_{j=1}^{M^A}\sum_{j'=1}^{M^A} \frac{(\pi_{jj'}^A - \pi_j^A \pi_{j'}^A)}{\pi_j^A \pi_{j'}^A} L_j^A L_{j'}^A. \tag{5.25}$$

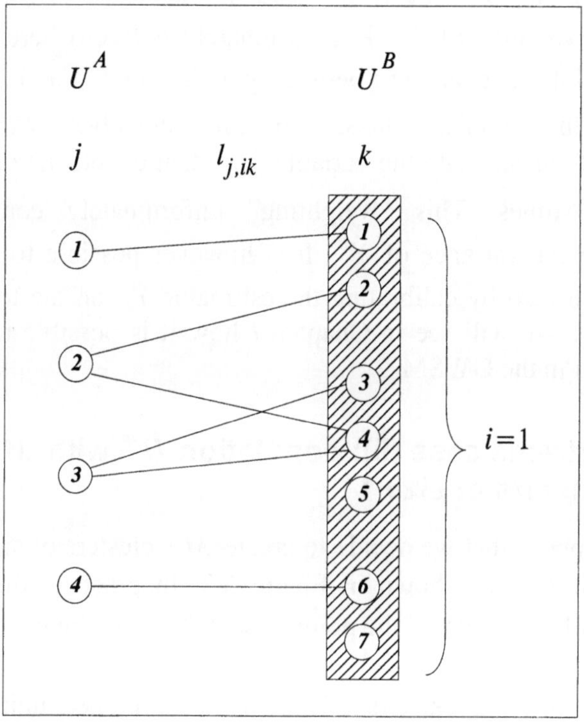

Figure 5.1: *Population U^B with a single cluster*

Looking at the variance formula (5.25), this variance will be zero if, for each unit j of the population U^A, the selection probability π_j^A is proportional to the total number of links L_j^A. In other words, it is preferable here to assign the selection probabilities of the units from the sampling frame U^A in such a way that they are proportional to the number of links coming from these units.

Note that the variance (5.25) also becomes zero in the case where there is only one link for each unit j of U^A (such as for the longitudinal surveys showed in Figure 3.1, for example) and where the sample s^A is selected by stratified simple random sampling.

However, the present discussion remains academic since we are in the extreme case where the population U^B has only a single cluster. In fact, by only choosing a single unit j of U^A, we should be able to estimate the total U^B with a zero variance owing to the fact that having only a single cluster leads to the census of the population U^B. The variance of $\hat{Y}^B$ is unfortunately not zero here due to the complex links that can exist between populations U^A and U^B.

With complex links, we find ourselves calculating a "weighted" mean (with the variable $l_{j,1k}$) that counts the census value Y^B many times. This "weighting" unfortunately contributes to increasing the variance of $\hat{Y}^B$. It is however possible to reduce this variance to zero by calibrating the estimator $\hat{Y}^B$ on the total number of links L. We will see in Chapter 7 how it is possible to introduce calibration in the GWSM.

5.2.2 Extreme case (ii): population U^B with M^B clusters of size 1

Suppose that we decide to create M^B clusters of size 1 for the population U^B, as shown in Figure 5.2. In practice, this case can cause problems in bias if the population U^B has unlinked units k, such as units 5 and 7 in Figure 2.1.

By forming clusters of size 1, the unlinked units will find themselves isolated from the population U^A and they will therefore not have any chance of being surveyed. Furthermore, note that this situation is directly contradicting Constraint 2.1 of the GWSM. In order to simplify the discussion, we will assume here that the population U^B does not have unlinked units to population U^A. In Figure 5.2, we thus added links between pairs (3,5) and (4,7) that we have represented by dotted lines.

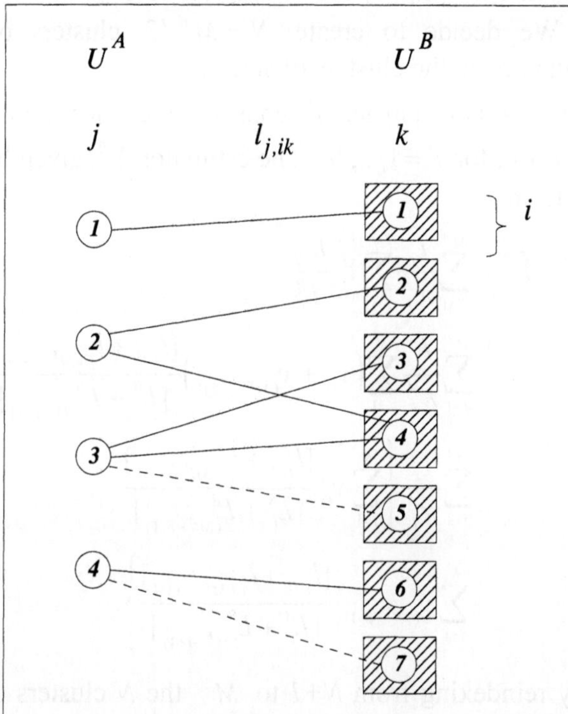

Figure 5.2: *Population U^B with clusters of size* 1

As in section 5.3.1, it is appropriate to use the form of $\hat{Y}^B$ given by Corollary 4.3. Since the target population U^B here has M^B clusters i of size $M_i^B = 1$, we have $Y_i = \sum_{k=1} y_{ik} = y_{i1}$, $L_{j,i} = \sum_{k=1} l_{j,ik} = l_{j,i1}$ and $L_i^B = \sum_{j=1}^{M^A} L_{j,i} = \sum_{j=1}^{M^A} l_{j,i1}$.

We then get

$$\hat{Y}^B = \sum_{j=1}^{M^A} \frac{t_j}{\pi_j^A} \sum_{i=1}^{M^B} Y_i \frac{L_{j,i}}{L_i^B} = \sum_{j=1}^{M^A} \frac{t_j}{\pi_j^A} \sum_{i=1}^{M^B} y_{i1} \frac{l_{j,i1}}{L_i^B}. \tag{5.26}$$

Unfortunately, the form (5.26) does not bring much information on the effect that the creation of clusters can have on the precision of the estimator $\hat{Y}^B$. This form, however, can be useful in studying the performance of $\hat{Y}^B$ if we create clusters of size 2.

In order to simplify the discussion, assume that M^B is an even number. We decide to create $N = M^B / 2$ clusters of size 2 by combining in pairs the clusters of size 1.

Let the new cluster i' consist of cluster $i = i'$ and cluster $i = M^B - i' + 1$, for $i' = 1,...,N$. The estimator $\hat{Y}^B$ given by (4.16) then takes the form

$$\begin{aligned}
\hat{Y}^B &= \sum_{j=1}^{M^A} \frac{t_j}{\pi_j^A} \sum_{i'=1}^{N} Y_{i'} \frac{L_{j,i'}}{L_{i'}^B} \\
&= \sum_{j=1}^{M^A} \frac{t_j}{\pi_j^A} \sum_{i=1}^{N} \left(y_{i1} + y_{(M^B-i+1)1} \right) \frac{[l_{j,i1} + l_{j,(M^B-i+1)1}]}{[L_i^B + L_{(M^B-i+1)}^B]} \\
&= \sum_{j=1}^{M^A} \frac{t_j}{\pi_j^A} \left\{ \sum_{i=1}^{N} y_{i1} \frac{[l_{j,i1} + l_{j,(M^B-i+1)1}]}{[L_i^B + L_{(M^B-i+1)}^B]} \right. \\
&\qquad \left. + \sum_{i=1}^{N} y_{(M^B-i+1)1} \frac{[l_{j,i1} + l_{j,(M^B-i+1)1}]}{[L_i^B + L_{(M^B-i+1)}^B]} \right\}.
\end{aligned} \tag{5.27}$$

By reindexing from *N+1* to M^B the *N* clusters of the second sum of (5.27), we can then write equation (5.27) under the form

$$\begin{aligned}
\hat{Y}^B &= \sum_{j=1}^{M^A} \frac{t_j}{\pi_j^A} \left\{ \sum_{i=1}^{N} y_{i1} \frac{[l_{j,i1} + l_{j,(M^B-i+1)1}]}{[L_i^B + L_{(M^B-i+1)}^B]} + \sum_{i=N+1}^{M^B} y_{i1} \frac{[l_{j,(M^B-i+1)1} + l_{j,i1}]}{[L_{(M^B-i+1)}^B + L_i^B]} \right\} \\
&= \sum_{j=1}^{M^A} \frac{t_j}{\pi_j^A} \left\{ \sum_{i=1}^{M^B} y_{i1} \frac{[l_{j,i1} + l_{j,(M^B-i+1)1}]}{[L_i^B + L_{(M^B-i+1)}^B]} \right\}.
\end{aligned} \tag{5.28}$$

By comparing equations (5.26) and (5.28), we can see the effect of forming clusters of size 2, compared to forming clusters of size 1. We first note that in the two equations, the variable of interest y_{i1} is "weighted" by a factor representing the ratio between the number of links for the cluster with unit *j* and the total number of links for the cluster. With the clusters of size 1, we see by equation (5.26) that each unit $i1$ is weighted by a factor dependent on a single link with unit *j*. With the clusters of size 2, equation (5.28) shows us that the factor then depends on two possible links with unit *j*. This factor proves to be decisive in the precision of the estimator $\hat{Y}^B$, as we can see in the following section.

5.2.3 General case and discussion

We will consider here the general case where each cluster i has M_i^B units. Again, we consider the form of the estimator $\hat{Y}^B$ given by Corollary 4.3. From (4.16), we have

$$\begin{aligned}\hat{Y}^B &= \sum_{j=1}^{M^A} \frac{t_j}{\pi_j^A} \sum_{i=1}^{N} Y_i \frac{L_{j,i}}{L_i^B} \\ &= \sum_{j=1}^{M^A} \frac{t_j}{\pi_j^A} \sum_{i=1}^{N} \sum_{k=1}^{M_i^B} y_{ik} \frac{L_{j,i}}{L_i^B}.\end{aligned} \tag{5.29}$$

We now use the following result.

Result 5.1

For all units j of the population U^A, we have

$$Z_j = \sum_{i=1}^{N} \sum_{k=1}^{M_i^B} y_{ik} \frac{L_{j,i}}{L_i^B}.$$

Proof

$$\begin{aligned}\sum_{i=1}^{N} \sum_{k=1}^{M_i^B} y_{ik} \frac{L_{j,i}}{L_i^B} &= \sum_{i=1}^{N} Y_i \frac{L_{j,i}}{L_i^B} = \sum_{i=1}^{N} \frac{Y_i}{L_i^B} \sum_{k=1}^{M_i^B} l_{j,ik} \\ &= \sum_{i=1}^{N} \sum_{k=1}^{M_i^B} l_{j,ik} \frac{Y_i}{L_i^B} = \sum_{i=1}^{N} \sum_{k=1}^{M_i^B} l_{j,ik} z_{ik} \\ &= Z_j.\end{aligned} \tag{5.30}$$

■

According to equation (5.29), in a general manner, each unit k of a cluster i of U^B is "weighted" by a factor representing the ratio between the number of links between cluster i and unit j, and the total number of links for cluster i. The larger the cluster size, the more we expect that the number of the links part of this factor will be large. As it is a question of the number of links for the whole cluster i containing unit k, the larger the cluster is, the more we expect that the factor "weighting" each unit ik has a kind of homogeneity. This homogeneity is important because of Result 5.1 that shows us that the double some coming from equation (5.29) is nothing more than the variable Z_j defined in Theorem 4.1. More important still is the fact that, by Corollary 4.2, the variable Z_j feeds directly into the calculation of the variance for the estimator $\hat{Y}^B$.

The homogeneity of the factor "weighting" each unit *ik* of U^B will contribute to the homogeneity of the variable Z_j. However, this homogeneity will only be favourable if the formation of the clusters is done in such a way that we combine the clusters that are already linked to the same unit *j* of U^A. Indeed, since the "weighting" of each unit *ik* of U^B depends on the number of links $L_{j,i}$ between cluster *i* and unit *j*, such a grouping of these clusters will not produce an increase in surveyed units following the selection of unit *j* of U^A. This can be seen from Result 5.1. For example, we decide to group cluster *i*=1 for which $L_{j,i} \neq 0$ and cluster *i*=2 for which $L_{j,i} = 0$. The new cluster i' will have $L_{j,i'} \neq 0$ and we will survey all units of this new cluster. As a consequence, we will now survey the units of cluster *i*=2 that were not included previously in the survey process because $L_{j,i} = 0$. We will thus have increased the size of Z_j and perhaps have created heterogeneity in this cluster, which could contribute to increasing the variance of $\hat{Y}^B$ instead of decreasing it.

We can also see this problem from the point of view of the estimation weights w_{ik} obtained by (2.4). By grouping clusters that are not linked to the same unit *j* of U^A, we risk combining clusters that would have been surveyed with other clusters that would not have been surveyed without this grouping. These new clusters would not have been surveyed simply because none of the units *j* of U^A to which they are linked would have been selected in s^A, i.e., that for a given cluster *i*, we would have $L_{j,i} \neq 0$ but $t_j = 0$. From (2.4), we see that the weights w_{ik} of the new clusters would only depend upon the selection probabilities π_j^A of the clusters that would have been surveyed before the grouping. We would then have large clusters, but whose weight would be calculated from some π_j^A only. Ernst (1989) noted this problem by pointing out that a use of a maximum number of selection probabilities π_j^A generally leads to an estimate of U^B with a larger precision. In section 4.6, we also saw that estimator (4.26) coming from the improvement of the estimator $\hat{Y}^B$ depends on the selection probability of the *n* clusters of Ω^B, and therefore on the selection probabilities of the units *j* de U^A having non-zero links with these *n* clusters. This estimator $\hat{Y}^{COND,B}$ (or $\hat{Y}^{RB,B}$) in fact has a variance less than or equal to the estimator $\hat{Y}^B$ because it uses the set of all selection probabilities π_j^A leading to the selection of the clusters of Ω^B, and not only the probabilities π_j^A of the units *j* selected in s^A.

It is important to note that if the links are complex, a grouping involving only the clusters that are linked to the same unit *j* of U^A is not always possible. Consider the example shown in Figure 5.2. We could consider grouping the clusters containing units 2 and 4 from U^B because they are linked to the same unit *j*=2 of U^A. However, this grouping will have an effect on the clusters linked to unit *j*=3, as this unit will now have an indirect link with unit 2 of U^B through unit 4.

In the case where the sampling design used for the selection of the sample s^A of U^A is of equal probabilities, Corollary 4.2 shows us that the homogeneity of the Z_j will contribute to reducing the variance. Thus, it seems that with this type of sampling design, we will have an advantage, for increasing the precision of the estimates, to form clusters of large size by combining as much as possible clusters that are already links to the same units *j* of U^A.

If the sampling design is of unequal probabilities, the variance will be zero if the variable Z_j is proportional to π_j^A for $j = 1,...,M^A$. If the links are complex, it is not clear how the clusters must be formed so that we have that proportionality. Indeed, as one unit *j* can also lead to surveying more than one cluster and that a same cluster can be surveyed due to the selection of more than one unit *j* of U^A, it is then necessary to control at the same time the formation of the clusters and the assignment of the selection probabilities π_j^A, which is very difficult to do in practice. If the links are not complex (being one-to-one, one-to-many, or many-to-one), it is possible to determine the selection probabilities π_j^A so that they are approximately proportional to the variables Z_j under the conditions, of course, of having an auxiliary variable correlated with the variable of interest *y* and of knowing the composition of the clusters *i* of U^B before surveying them.

Example 5.1

As an example, consider the case shown in Figure 5.2.

Units of U^B from Fig. 5.2	i	k	z_{ik}		
1	1	1	$y_{11}/1$	Z_1	$\frac{y_{11}}{1}$
2	2	1	$y_{21}/1$	Z_2	$\frac{y_{21}}{1}+\frac{y_{41}}{2}$
3	3	1	$y_{31}/1$		
4	4	1	$y_{41}/2$	Z_3	$\frac{y_{31}}{1}+\frac{y_{41}}{2}+\frac{y_{51}}{1}$
5	5	1	$y_{51}/1$		
6	6	1	$y_{61}/1$	Z_4	$\frac{y_{61}}{1}+\frac{y_{71}}{1}$
7	7	1	$y_{71}/1$		

If we select from U^A unit j=1 and unit j=2, the estimator $\hat{Y}^B$ is then written

$$\hat{Y}^B = \frac{1}{\pi_1^A}(y_{11}) + \frac{1}{\pi_2^A}\left(y_{21} + \frac{y_{41}}{2}\right).$$

Assuming that this sample was selected with a simple random sample of $m^A = 2$ units chosen among $M^A = 4$, we have $\pi_j^A = \pi^A = 1/2$. This estimator becomes

$$\hat{Y}^B = 2 \times \left(y_{11} + y_{21} + \frac{y_{41}}{2}\right).$$

Moreover, assuming that we measure the value $y_{ik} = 1$ for all units surveyed in U^B with the goal of estimating M^B, $\hat{Y}^B = \hat{M}^B = 2(1+1+1/2) = 5$. The variance of $\hat{M}^B$ is given by

$$Var(\hat{M}^B) = \frac{(M^A)^2}{m^A}\left(1 - \frac{m^A}{M^A}\right)\frac{1}{M^A - 1}\left(\sum_{j=1}^{M^A} Z_j^2 - \frac{(\sum_{j=1}^{M^A} Z_j)^2}{M^A}\right)$$

$$= \frac{16}{2}\left(1 - \frac{2}{4}\right)\frac{1}{3}\left(\sum_{j=1}^{4} Z_j^2 - \frac{49}{4}\right)$$

$$= \frac{4}{3}\left(1 + \frac{9}{4} + \frac{25}{4} + 4 - \frac{49}{4}\right) = \frac{5}{3} = 1.6667.$$

Suppose that we combine the adjacent cluster pairs to form new clusters. We then combine clusters *i*=1 and *i*=2, *i*=3 and *i*=4, as well as *i*=5 and *i*=6. Then, we can calculate, as presented in the following table, the new values of Z_j.

Units of U^B from Fig. 5.2	i'	k	$z_{i'k}$		
1	1	1	$\frac{y_{11}+y_{12}}{2}$	Z_1	$\frac{y_{11}+y_{12}}{2}$
2	1	2	$\frac{y_{11}+y_{12}}{2}$	Z_2	$\frac{y_{11}+y_{12}}{2}+\frac{y_{21}+y_{22}}{3}$
3	2	1	$\frac{y_{21}+y_{22}}{3}$		
4	2	2	$\frac{y_{21}+y_{22}}{3}$	Z_3	$2\times\frac{y_{21}+y_{22}}{3}+\frac{y_{31}+y_{32}}{2}$
5	3	1	$\frac{y_{31}+y_{32}}{2}$		
6	3	2	$\frac{y_{31}+y_{32}}{2}$	Z_4	$\frac{y_{31}+y_{32}}{2}+\frac{y_{41}}{1}$
7	4	1	$\frac{y_{41}}{1}$		

Selecting unit *j*=1 and unit *j*=2, the estimator $\hat{Y}^B$ can then be written

$$\hat{Y}^B = \frac{1}{\pi_1^A}\left(\frac{y_{11}+y_{12}}{2}\right)+\frac{1}{\pi_2^A}\left(\frac{y_{11}+y_{12}}{2}+\frac{y_{21}+y_{22}}{3}\right).$$

If this sample was selected with a simple random sample of $m^A = 2$ units chosen among $M^A = 4$, we have

$$\hat{Y}^B = 2\times\left(2\times\frac{y_{11}+y_{12}}{2}+\frac{y_{21}+y_{22}}{3}\right) = 2\times\left(y_{11}+y_{12}+\frac{y_{21}+y_{22}}{3}\right).$$

Furthermore, assuming that we measure the value $y_{ik}=1$ for all units surveyed in U^B, we can conclude that $\hat{Y}^B = \hat{M}^B = 2(1+1+2/3) = 16/3 = 5.3333$. The variance of $\hat{M}^B$ is given by

$$Var(\hat{M}^B) = \frac{4}{3}\left(\sum_{j=1}^{4} Z_j^2 - \frac{49}{4}\right) = \frac{4}{3}\left(1 + \frac{25}{9} + \frac{49}{9} + 4 - \frac{49}{4}\right) = 1.2962.$$

Suppose that we again combine adjacent clusters into pairs to form new clusters. As for the previous table, we combine in this manner clusters $i' = 1$ and $i' = 2$, as well as $i' = 3$ and $i' = 4$. We can then calculate the new values of Z_j.

Units of U^B from Fig. 5.2	i''	k	$z_{i''k}$		
1	1	1	$\frac{y_{11}+y_{12}+y_{13}+y_{14}}{5}$	Z_1	$\frac{y_{11}+y_{12}+y_{13}+y_{14}}{5}$
2	1	2	$\frac{y_{11}+y_{12}+y_{13}+y_{14}}{5}$	Z_2	$2\times\frac{y_{11}+y_{12}+y_{13}+y_{14}}{5}$
3	1	3	$\frac{y_{11}+y_{12}+y_{13}+y_{14}}{5}$		
4	1	4	$\frac{y_{11}+y_{12}+y_{13}+y_{14}}{5}$	Z_3	$2\times\frac{y_{11}+y_{12}+y_{13}+y_{14}}{5} + \frac{y_{21}+y_{22}+y_{23}}{3}$
5	2	1	$\frac{y_{21}+y_{22}+y_{23}}{3}$		
6	2	2	$\frac{y_{21}+y_{22}+y_{23}}{3}$	Z_4	$2\times\frac{y_{21}+y_{22}+y_{23}}{3}$
7	2	3	$\frac{y_{21}+y_{22}+y_{23}}{3}$		

Selecting unit j=1 and unit j=2, the estimator $\hat{Y}^B$ is then written

$$\hat{Y}^B = \frac{1}{\pi_1^A}\left(\frac{y_{11}+y_{12}+y_{13}+y_{14}}{5}\right) + \frac{2}{\pi_2^A}\left(\frac{y_{11}+y_{12}+y_{13}+y_{14}}{5}\right).$$

Selecting this sample with a simple random sample of $m^A = 2$ units chosen among $M^A = 4$, we have

$$\hat{Y}^B = 6\times\left(\frac{y_{11}+y_{12}+y_{13}+y_{14}}{5}\right).$$

Furthermore, assuming that we measure the value $y_{ik} = 1$, $\hat{Y}^B = \hat{M}^B = 6(4/5) = 24/5 = 4.8$. The variance of $\hat{M}^B$ is given by

$$Var(\hat{M}^B) = \frac{4}{3}\left(\sum_{j=1}^{4} Z_j^2 - \frac{49}{4}\right) = \frac{4}{3}\left(\frac{16}{25} + \frac{64}{25} + \frac{169}{25} + 4 - \frac{49}{4}\right) = 2.28 .$$

Finally, suppose that we combine the two clusters from the previous table to form a single cluster. We can then calculate the new values of Z_j.

Units of U^B from Fig. 5.2	i'''	k	$z_{i'''k}$		
1	1	1	$\frac{\sum_{k=1}^{7} y_{1k}}{8}$	Z_1	$\frac{\sum_{k=1}^{7} y_{1k}}{8}$
2	1	2	$\frac{\sum_{k=1}^{7} y_{1k}}{8}$	Z_2	$2 \times \frac{\sum_{k=1}^{7} y_{1k}}{8}$
3	1	3	$\frac{\sum_{k=1}^{7} y_{1k}}{8}$		
4	1	4	$\frac{\sum_{k=1}^{7} y_{1k}}{8}$	Z_3	$3 \times \frac{\sum_{k=1}^{7} y_{1k}}{8}$
5	1	5	$\frac{\sum_{k=1}^{7} y_{1k}}{8}$		
6	1	6	$\frac{\sum_{k=1}^{7} y_{1k}}{8}$	Z_4	$2 \times \frac{\sum_{k=1}^{7} y_{1k}}{8}$
7	1	7	$\frac{\sum_{k=1}^{7} y_{1k}}{8}$		

Selecting units *j*=1 and *j*=2, the estimator $\hat{Y}^B$ is then written

$$\hat{Y}^B = \frac{1}{\pi_1^A}\left(\frac{\sum_{k=1}^{7} y_{1k}}{8}\right) + \frac{4}{\pi_2^A}\left(\frac{\sum_{k=1}^{7} y_{1k}}{8}\right).$$

Selecting this sample with a simple random sample of $m^A = 2$ units chosen among $M^A = 4$, we have

$$\hat{Y}^B = 5 \times \left(\frac{\sum_{k=1}^{7} y_{1k}}{4}\right).$$

Finally, supposing moreover that we measure the value $y_{ik} = 1$, $\hat{Y}^B = \hat{M}^B = 5(7/4) = 8.75$. The variance of $\hat{M}^B$ is given by

$$Var(\hat{M}^B) = \frac{4}{3}\left(\sum_{j=1}^{4} Z_j^2 - \frac{49}{4}\right) = \frac{4}{3}\left(\frac{49}{64} + \frac{196}{64} + \frac{441}{64} + \frac{196}{64} - \frac{49}{4}\right) = 2.0417.$$

In addition to the considerations of precision of the estimates, there are operational reasons that themselves encourage us to not form large clusters. The first reason is the difficulty of creating the list of links for the selected clusters. As we can see in steps 2 and 3 of the GWSM given in Chapter 2, the use of the GWSM requires us to know the total number of links L_{ik}^B for each unit k of the clusters i for Ω^B. This is necessary to get the total number of links $L_i^B = \sum_{k=1}^{M_i^B} L_{ik}^B$ for each cluster i of Ω^B. If the cluster is large, this quantity can be difficult, indeed even impossible, to establish in practice. In the case where the two populations U^A and U^B are lists where we know all the links $l_{j,ik}$ between units j of U^A and units ik of U^B, this does not pose a problem. In the case of social surveys, however, the compositions of the households are often established during the interviews themselves. If the clusters are no longer the households but rather a much larger entity (the neighbourhood, for example), it will then be much more difficult to establish the number of links for the clusters of Ω^B. Ardilly and Le Blanc (1999) as well as Ardilly and Le Blanc (2001) noted this problem during the use of the GWSM for the weighting of a survey of homeless people. Section 8.7 will deal with the problem of links identification.

The second operational reason to not form large clusters is related to the instability of collection costs. Recall that following the selection of units j from s^A, we identify the units ik of U^B that have a non-zero link with these units j, and we finally go and survey all units of the clusters i containing the identified units ik. Therefore, the selection of each unit j from U^A leads to the surveying of an entire cluster. If this cluster is large, we will then have an imposing collection cost associated with each cluster. In the case where the links between U^A and U^B are one-to-one or many-to-one, the m^A units of s^A will be linked to at most m^A clusters of U^B. We can then control the maximum collection cost. In the case where certain links are one-to-many, there will be a large instability in the collection costs. Indeed, by selecting one unit j linked to a single cluster, we will have the collection costs of this single cluster. On the other hand, by selecting one unit j linked to two clusters, we will then double the collection costs, and so on if unit j is linked to more clusters. We easily see that if the clusters are small, these variations in collection costs can be negligible. In the opposite, if the clusters are large, they can cause enormous budgetary problems.

The instability of collection costs can also be the result of a large disparity between the sizes of clusters. By allowing for the creation of large clusters, we at the same time allow a much larger variability between cluster sizes. For example, in social surveys, by extending the clusters to the neighbourhood level instead of the household, these new clusters will be of variable size if the neighbourhoods are not all of the same size. The variability in the size of the neighbourhoods is generally much larger than that of the households, as neighbourhoods can contain between hundreds or even several thousands of people, whereas households contain, most often, between one to five people. If the selection of different units j from U^A leads to the surveying of clusters of very variable size, it will then be very difficult to control the collection costs. To better control these costs, it will thus be worthwhile to form small clusters of relatively equal size.

5.3 ELIMINATION OF THE NOTION OF CLUSTERS

According to the survey process, a sample s^A from U^A is selected that leads to the identification of n clusters from U^B. For each of these clusters i, all M_i^B units contained in the clusters are then surveyed. The survey process is therefore performed in two steps. From Figure 2.1, we can illustrate these two steps with Figure 5.3.

We can also see this process as having a single step. A sample s^A from U^A is selected that then leads directly to the identification of m^B units from U^B, where $m^B = \sum_{i=1}^{n} M_i^B$. This way of seeing the process eliminates the notion of clusters used up to now. However, it requires extending the structure of the links in such a way that a unit j from U^A that had a non-zero link with a unit k of a cluster i from U^B now has non-zero links with all units k of this cluster i.

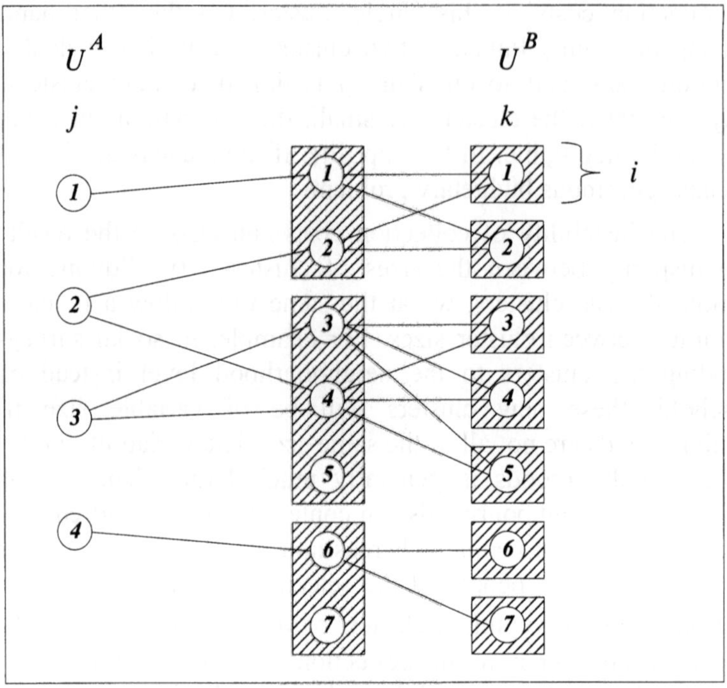

Figure 5.3: *Survey process seen in two steps*

One way of extending the structure of the links for the clusters is to define a new indicator variable $l^*_{j,ik}$ to identify the links between units *j* of U^A and units *ik* of U^B. We then define $l^*_{j,ik} = 1$ for each unit $k \in U^B_i$ if $l_{j,ik} = 1$ for at least one unit $k \in U^B_i$, and 0 otherwise. In other words, $l^*_{j,ik} = 1 - \prod_{j=1}^{M^B_i} (1 - l_{j,ik})$. From Figure 2.1 (or Figure 5.3), we then get Figure 5.4. Note that a similar structure of links will be used in section 6.5 in the context of longitudinal surveys.

By applying steps 1 to 4 of the GWSM, we can obtain the estimation weight w^*_{ik} by replacing the indicator variable $l_{j,ik}$ with the new variable $l^*_{j,ik}$. However, note that the resulting estimation weight w^*_{ik} is different from the estimation weight w_{ik} obtained by the GWSM with the indicator variable $l_{j,ik}$. This can be illustrated by the small example that we present in the two Figures 5.5.

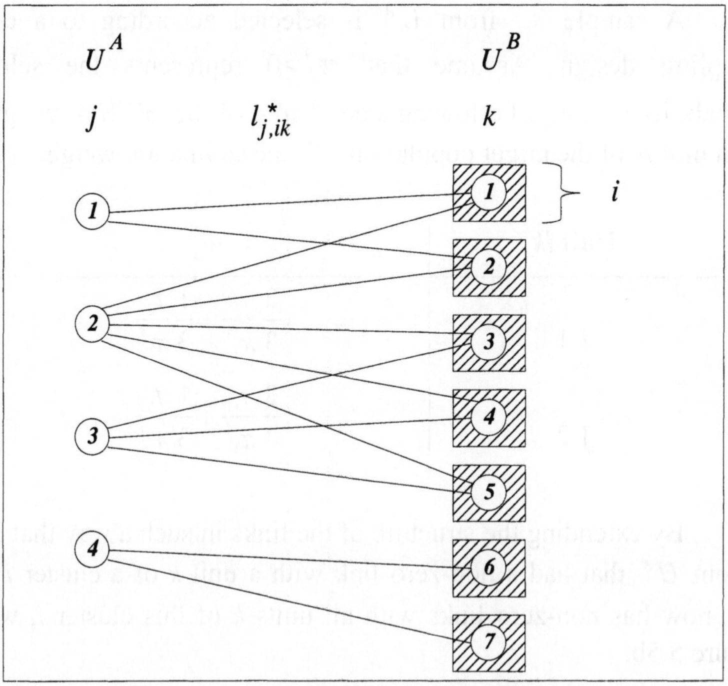

Figure 5.4: *Example of extending the structure of links*

Let the populations U^A and U^B be represented in Figure 5.5a. The population U^B only has a single cluster of size 2.

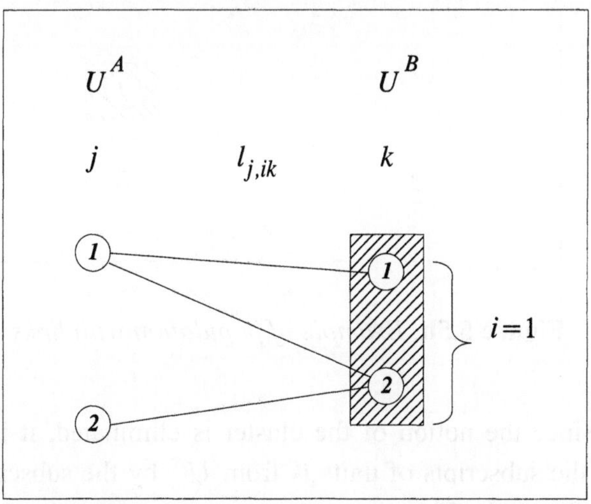

Figure 5.5a: *Example of population with links* $l_{j,ik}$

A sample s^A from U^A is selected according to a certain sampling design. Assume that $\pi_j^A>0$ represents the selection probability of unit *j*. Following steps 1 to 4 of the GWSM, we get for each unit *ik* of the target population U^B the estimation weight w_{ik}.

Unit *ik*	w_{ik}
1 1	$\frac{2}{3}\frac{t_1}{\pi_1^A}+\frac{1}{3}\frac{t_2}{\pi_2^A}$
1 2	$\frac{2}{3}\frac{t_1}{\pi_1^A}+\frac{1}{3}\frac{t_2}{\pi_2^A}$

By extending the structure of the links in such a way that a unit *j* from U^A that had a non-zero link with a unit *k* of a cluster *i* from U^B now has non-zero links with all units *k* of this cluster *i*, we get Figure 5.5b.

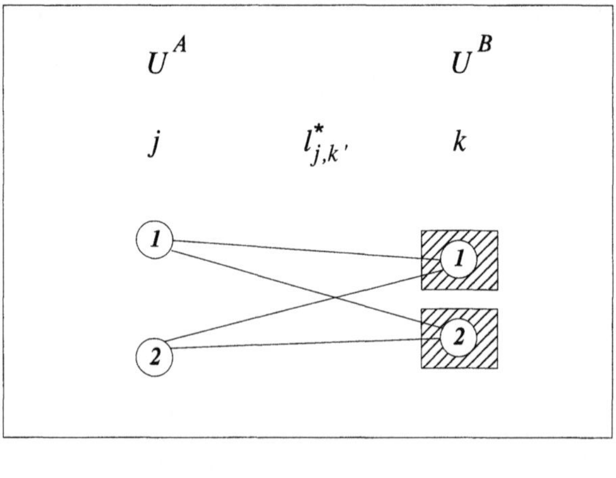

Figure 5.5b: *Example of population with links* $l^*_{j,k'}$

Since the notion of the cluster is eliminated, it is practical to replace the subscripts of units *ik* from U^B by the subscript k' where $k'=k+\sum_{i'=1}^{i-1}M_{i'}^B$. This new subscript no longer uses the subscript *i* linked to the clusters. By using the new indicator variable $l^*_{j,k'}$ (or $l^*_{j,ik}$) in steps 1 to 4 of the GWSM, we get the following estimation weights $w^*_{k'}$:

Unit k'	$w^*_{k'}$
1	$\frac{1}{2}\frac{t_1}{\pi_1^A}+\frac{1}{2}\frac{t_2}{\pi_2^A}$
2	$\frac{1}{2}\frac{t_1}{\pi_1^A}+\frac{1}{2}\frac{t_2}{\pi_2^A}$

It is possible to construct the estimation weights after eliminating the notion of the cluster so that they are the same as before the elimination. To do this, we use the weighted links described in section 4.5. Starting with (2.5), it is sufficient to set $\widetilde{\theta}_{j,ik} = L_{j,i} / L_i^B$ for all units k from the clusters i of U^B. Note that this definition of the constants $\widetilde{\theta}_{j,ik}$ must be taken **before** extending the links. We proceed subsequently by extending the links in such a way that a unit j from U^A that had a non-zero link with a unit k of a cluster i from U^B now has non-zero links with all units k of this cluster i. Again, starting from Figure 2.1, we get Figure 5.4. We then replace the subscripts of units ik from U^B by the subscript k'. We then have a value $\widetilde{\theta}_{j,k'}$ for each unit k' of the target population U^B (without the notion of the cluster).

Following steps 1 to 3 of the weighted version of the GWSM given in section 4.5, we get the following estimation weight:

$$\begin{aligned} w^{\theta}_{k'} &= \sum_{j=1}^{M^A} \widetilde{\theta}_{j,k'} \frac{t_j}{\pi_j^A} \\ &= \sum_{j=1}^{M^A} \frac{t_j}{\pi_j^A} \frac{L_{j,i}}{L_i^B} \end{aligned} \tag{5.31}$$

for $k' \in U_i^B$, the old cluster i.

By comparing (5.31) and (2.5), we see that the estimation weight $w_{k'}^{\theta}$ of unit k' from the population U^B without clusters is the same as the estimation weight w_{ik} of unit k from cluster i obtained from the population U^B with clusters. We can illustrate this result by again using the small example from Figure 5.5. For the units j from U^A and k' from U^B of Figure 5.5b, we first of all get the following values of $\widetilde{\theta}_{j,k'}$:

Unit *j* of U^A	**Unit *k'* of U^B**	$\widetilde{\theta}_{j,k'}$
1	1	$\frac{2}{3}$
	2	$\frac{2}{3}$
2	1	$\frac{1}{3}$
	2	$\frac{1}{3}$

From (5.31), we then get the following estimation weights:

Unit *k'*	$w_{k'}^{\theta}$
1	$\frac{2}{3}\frac{t_1}{\pi_1^A}+\frac{1}{3}\frac{t_2}{\pi_2^A}$
2	$\frac{2}{3}\frac{t_1}{\pi_1^A}+\frac{1}{3}\frac{t_2}{\pi_2^A}$

The weights $w_{k'}^{\alpha}$ are quite comparable to those obtained by the GWSM with the notion of the cluster.

As seen, it is possible to eliminate the notion of the cluster for the GWSM. To do this, it is at first sufficient to extend the links in such a way that a unit j from U^A that had a non-zero link with a unit k of a cluster i from U^B now has non-zero links with all units k of this cluster i. Secondly, to get the same estimation weights as the GWSM with the clusters, the weighted version of the GWSM is used by setting $\widetilde{\theta}_{j,ik} = L_{j,i} / L_i^B$.

Although it can be interesting to eliminate the notion of the cluster for the GWSM, it is more natural to work with the clusters. Indeed, as mentioned in section 5.2, the clusters are most of the time formed in a natural manner. In social surveys, they generally correspond to households, whereas in economic surveys they often correspond to enterprises. Eliminating the notion of the cluster also contradicts the recommended survey process that is the one which surveys all units of the clusters identified by the units j of s^A. Recall that this allows for savings in collection costs and allows us to produce estimates at the cluster level.

The extension of the links also goes against the nature of the problems treated. The links represent a certain connection between the two populations U^A and U^B. The fact that units from the target population U^B do not have a link with the population U^A often refers to a natural process. For example, at the level of longitudinal surveys, individuals that belong to a household (or cluster) from population U^B and that do not have a link with U^A are either immigrants within the population or newborns. We will further discuss in detail this problem in Chapter 6. At the level of economic surveys, if we refer to Figure 1.3, the establishments unlinked to enterprises (or clusters) are establishments absent from the sampling frame. The extension of the links to the set of units from the clusters contributes toward hiding this aspect of the problem. It is worth noting that the extension of the links is, anyway, performed during the application of the GWSM, but this is done in an implicit manner.

Finally, the elimination of the notion of the cluster remains artificial since we must obtain the constants $\widetilde{\theta}_{j,ik}$ to apply the weighted version of the GWSM. Recall that the constants $\widetilde{\theta}_{j,ik}$ depend on the number $L_{j,i}$ of links between cluster i from U^B and unit j from U^A, as well as the number L_i^B of links for cluster i from U^B. As a result, we cannot completely eliminate the notion of the cluster.

CHAPTER 6

APPLICATION IN LONGITUDINAL SURVEYS

Longitudinal surveys, i.e., surveys that follow units over time, are steadily gaining importance within statistical agencies. Statistics Canada currently has three major longitudinal surveys of individuals: the National Population Health Survey, the National Longitudinal Survey of Children, and the Survey of Labour and Income Dynamics (SLID).

The primary objective of these surveys is to obtain longitudinal data. One of the uses of these data is to study the changes in variables over time (for example, longitudinal data may be used to analyse the chronic aspect of poverty).

A secondary objective is the production of *cross-sectional estimates*, in other words, estimates that represent the population at a given point in time. Although these estimates are far less important than the longitudinal data, to many users they are an essential aspect of the survey. Obtaining a representative cross-sectional view of the current population can be found to be useful for measuring the evolution of the population over time. The longitudinal aspect of the survey also improves the accuracy of the measurement of change.

We propose to apply the GWSM to longitudinal surveys and, in particular, to SLID. In the context of longitudinal surveys, the sampling frame U^A can be associated to the initial population (wave 1), while the target population U^B is the population a few years later (which will be called wave 2).

The GWSM is used here so that longitudinal samples can be used for cross-section estimation. The difficulty arises from the fact that, although the longitudinal sample remains constant, the distribution of the population (individuals and households) changes over time. At the individual level, these changes are produced by such events as births and deaths, immigration and emigration, and moves from one place to another. Obviously, the birth or death of an individual also changes household composition. Events such as marriage, divorce, separation, departure of a child and cohabitation are all factors that affect the population distribution within the household. If we are to obtain accurate and unbiased cross-sectional estimates based on a longitudinal sample, we need an estimation method that takes these changes into account.

As seen in Chapter 3, the fair share method and the weight share method (the precursor to the GWSM) were already used in the context of longitudinal surveys. This was described by Huang (1984), Judkins *et al.* (1984), Ernst, Hubble and Judkins (1984), and Ernst (1989).

The use of the GWSM instead of these methods, however, allows us to establish a more general theory, which leads to, among others, a simple variance calculation for the estimates (Lavallée, 1995). Deville (1998a) also discussed the GWSM in the context of longitudinal surveys.

Note that other methods, different from the weight share method and the GWSM, were studied to perform the weighting of longitudinal surveys and, in particular, for SLID. Lavallée and Hunter (1993) as well as Gailly and Lavallée (1993) considered the use of a composite (or combined) estimator where the sampled units are weighted differently depending on whether or not they are part of the longitudinal sample. Their research showed that the GWSM produces estimates with variances equal to those of the composite estimator, but the GWSM has the advantage of producing unbiased estimates.

6.1 SAMPLING DESIGN OF SLID

In January 1994, SLID was launched by Statistics Canada. Its aim is to observe individual activity in the labour market over time and changes in individual income and family circumstances. SLID first and foremost provides longitudinal data. However, cross-sectional estimates are also produced.

The target population of SLID is all persons, with no distinction as to age, who live in the provinces of Canada. For operational reasons, the Territories, institutions, Indian reserves, and military camps are excluded. For more details, see Lavallée (1993), Lavigne and Michaud (1998), as well as Lévesque and Franklin (2000).

6.1.1 Initial sample

The SLID longitudinal sample was drawn in January 1993. Although selected in January 1993, the survey formally began in January 1994; the January 1993 survey in fact served in obtaining preliminary data on the longitudinal individuals. This first panel of longitudinal individuals was surveyed for a period of six years, in addition to the preliminary interview. Thus, this panel selected in January 1993 was surveyed from 1994 to 1999. Note that a second panel of the same type was selected in January 1996 and was surveyed from 1997 to 2002. At the end of the first panel, a third panel selected in January 1999 was set up in order to replace the first one. This use of "superimposed" panels allows for different longitudinal samples starting in different years to be obtained. The panel rotation design is illustrated in Table 6.1, taken from Lavigne and Michaud (1998). For the current discussion, we will limit ourselves to the first panel selected in January 1993.

Table 6.1: *Panel rotation in SLID*

Panel	**Years**													
	93	94	95	96	97	98	99	00	01	02	03	04	05	06
1	P	I	I	I	I	I	I							
2				P	I	I	I	I	I	I				
3							P	I	I	I	I	I	I	
4										P	I	I	I	I

P: Preliminary interviews **I**: Interview on labour and income

The initial sample (or first panel) of SLID comes from two groups rotating out of the Canadian Labour Force Survey (LFS), making the sample a subsample of the LFS. The longitudinal sample for SLID is made up of close to 15,000 households. A *household* is defined as any person or group of persons living in a dwelling. It may consist of one person living alone, a group of people who are not related but who share the same dwelling, or the members of a family.

LFS is a periodic survey designed to produce monthly estimates of employment, self-employment and unemployment. This survey uses a stratified multi-stage sampling design that uses an area frame in which dwellings are the final sampling units. All the individuals who are members of households that occupy the selected dwellings make up the LFS sample. In other words, LFS draws a sample of dwellings and all individuals in the households that live in the selected dwellings are surveyed. A six-group rotation design is used to construct the sample: every month, one group that has been in the sample for six months is rotated out. Each rotation group contains approximately 10,000 households, or approximately 20,000 individuals 16 years of age or older. For further details on the LFS sample design, see Singh *et al.* (1990) and Dufour *et al.* (1998).

The longitudinal sample for SLID is not updated following its selection in January 1993. However, to give the sample some cross-sectional representativeness, *initially-absent individuals* in the population (i.e., individuals who were not part of the population in the year the longitudinal sample was selected) are considered in the sample in January 1994 and later. Initially-absent individuals include *newborns* (births since January 1993) and *immigrants*. Note that this addition to the sample is cross-sectional in that only the longitudinal individuals are permanently included in the sample.

Table 6.2 presents the terminology developed for SLID.

Table 6.2: *SLID Terminology*

Individuals
Longitudinal individuals: Individuals selected at wave 1 in the longitudinal sample.
Initially-absent individuals: Individuals who were not part of the population in the year the longitudinal sample was selected (wave 1). It includes immigrants and newborns.
Initially-present individuals: Individuals who were part of the population of wave 1 but were not selected then.
Cohabitants: Initially-absent and initially-present individuals who join a longitudinal household.
Immigrants: Individuals who, in January of wave 1, were outside the ten provinces of Canada and individuals who live in excluded areas (the Territories, institutions, Indian reserves and military bases).
Newborns: Births since January of wave 1.
Households
Longitudinal households: Households containing at least one longitudinal individual.

After sample selection in January 1993 (wave 1), the population contains longitudinal individuals and initially-present individuals. In January 1994 (wave 2), for example, the population contains longitudinal individuals, initially-present individuals and initially-absent individuals. Focusing on the households containing at least one longitudinal individual (i.e., *longitudinal households*), initially-present and initially-absent individuals who join these households are referred to as *cohabitants*.

SLID follows individual and household characteristics over time. At the time of each wave of interviews, all the members of a longitudinal household are surveyed. The composition of the longitudinal households changes over time, as the result of a birth or the arrival of an immigrant in the household. A part of the selection of initially-absent individuals is based on individuals who join longitudinal households.

6.1.2 Supplementary sample

Restricting the selection of initially-absent individuals who join longitudinal households unfortunately excludes households made up of initially-absent individuals only (for example, families of immigrants). To offset this shortcoming, one possibility is to select a *supplementary sample*. For example, this sample could be one of dwellings drawn directly from the ongoing LFS at each wave of interviews. Supplementary questions can then be added to the LFS questionnaire to detect households that contain **at least one immigrant**; the households selected are then surveyed.

Recalling that the supplementary sample is used for the selection of households made up solely of initially-absent individuals (i.e., immigrants and newborns), restricting this sample to immigrants does not pose any problem in representativeness. This is because it is highly unlikely that households containing only newborns would be found; each household normally contains at least one adult. The newborns are then already represented in the sample by the longitudinal households. Now, if the supplementary sample were to include newborns in addition to immigrants, significant costs would be added to the survey. This is because the supplementary sample would include a complete household for each newborn selected, producing excessive sample growth and unnecessary collection costs since the newborns are already represented in the initial sample.

Instead of using the ongoing LFS, another different approach is to select the supplementary sample by revisiting the dwellings used for the selection of the initial sample. This method offers some practical advantages (for example, it is easier to go to known addresses). This approach, however, brings the problem of new dwellings which were not there in January 1993. These dwellings have a zero probability of being selected in the supplementary sample, which introduces a source of bias. This is one reason why we favour the first approach, i.e., detecting households that contain at least one immigrant via the questionnaire of the LFS.

Figure 6.1 summarises the longitudinal and cross-sectional selection of individuals.

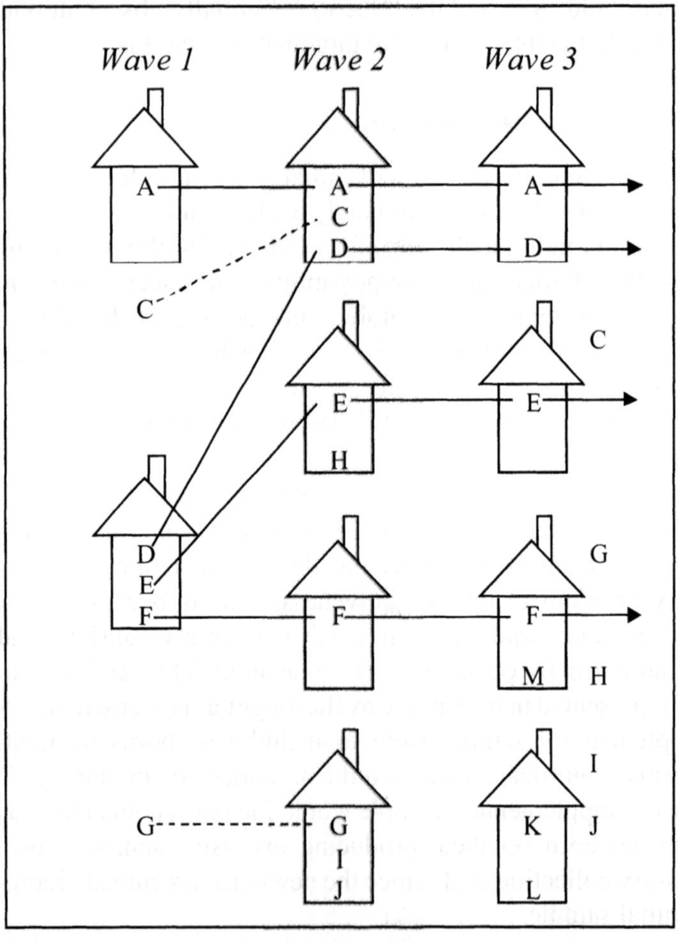

Figure 6.1: *Example of the selection of individuals within households*

In Figure 6.1, the letters and houses represent individuals and households, respectively. Individuals **A, D, E,** and **F** are longitudinal individuals whom we follow over time. Individual **C** is an initially-present individual, i.e., an individual who was included in the population in wave 1 but was not selected then. Initially-absent and initially-present individuals who join a longitudinal household are called cohabitants. In wave 2, individual **H** represents an initially-absent individual who joins the sample as a cohabitant.

The fourth house in wave 2 represents a household selected for the supplementary sample of wave 2 and in which individuals **I** and **J** are initially-absent individuals (with one of the two being necessarily an immigrant since the supplementary sample is restricted to them). Individual **G** is an initially-present individual with the same status as **C**. In wave 3, individuals **C** and **H** have left their longitudinal households and will therefore not be surveyed. Individuals **I** and **J** who were selected in the supplementary sample are now replaced with the individuals of the supplementary sample of wave 3, i.e., individuals **K** and **L**. Individual **M** is an initially-absent individual joining a longitudinal household as a cohabitant. It may finally be noted that, for cross-sectional representativeness, a selected household may contain one or more longitudinal individuals, initially-present individuals and initially-absent individuals (newborns and immigrants).

6.2 ESTIMATION WEIGHTS

To produce cross-sectional estimates, the longitudinal sample augmented with initially-absent individuals and initially-present individuals must be weighted. So, we look to obtain an estimation weight for each individual in each surveyed household. Note that the estimation weight of which we speak here is that before any adjustment for non-response and calibration (or post-stratification). It is, so to speak, the equivalent of the sample weight. It should be noted that the estimation weights here are useful solely for cross-sectional estimation.

The estimation weights are obtained from the selection probabilities. As mentioned above, in January 1993 (wave 1), we select for SLID a sample $s^{(1)}$ of $m^{(1)}$ individuals from a population $U^{(1)}$ of $M^{(1)}$ individuals. The sample is selected through dwellings which contain households. In other words, the $m^{(1)}$ individuals are obtained by selecting $n^{(1)}$ households from $N^{(1)}$, each household ι having a selection probability $\pi_{\iota}^{(1)} > 0$, $\iota = 1,..., N^{(1)}$. Let $M_{\iota}^{(1)}$ be the size of household ι

so that $M^{(1)} = \sum_{\imath=1}^{N^{(1)}} M_\imath^{(1)}$. Also let $\pi_{\imath j}^{(1)}$ be the selection probability of individual j from household $\imath$. We have $\pi_{\imath j}^{(1)} = \pi_\imath^{(1)}$ for all individuals j of household $\imath$. This selection probability is retained throughout all waves of the survey. In order to simplify the notation, we will omit the subscript $\imath$ related to the households and thus write $\pi_j^{(1)}$.

For a given subsequent wave (which may be defined as wave 2), the population U contains the $M^{(1)}$ individuals present at wave 1, plus $M^{(2)}$ initially-absent individuals (i.e., initially absent from the population at wave 1). The initially-absent individuals are immigrants or newborns. The population of initially-absent individuals is indicated by $U^{(2)}$. Hence, the population $U = U^{(1)} \cup U^{(2)}$ contains $M = M^{(1)} + M^{(2)}$ individuals. Letting $U^{*(2)}$ be the population of $M^{*(2)}$ immigrants (i.e., excluding newborns) of wave 2, we have $U^{*(2)} \subseteq U^{(2)}$, and also $M^{*(2)} \leq M^{(2)}$. In our notation, the asterisk (*) is used to specify that the newborns have been excluded. The individuals of wave 2 are contained in N households where household i is of size M_i, i=1,..., N.

For cross-sectional representativeness, some immigrants are selected from the supplementary sample. At wave 2, we then select a sample $s^{*(2)}$ of $m^{*(2)}$ immigrants from the population $U^{*(2)}$ of $M^{*(2)}$ immigrants. The $m^{*(2)}$ individuals from the supplementary sample are obtained by selecting $n^{*(2)}$ households from $N^{*(2)}$ where $N^{*(2)}$ represents the number of households from $U^{*(2)}$ containing at least one immigrant. The selection probability of household $\imath$ is given by $\pi_\imath^{*(2)}$ where we assume that $\pi_\imath^{*(2)} > 0$ for $\imath = 1,...,N^{*(2)}$. Let $\pi_{\imath j}^{*(2)}$ be the selection probability of immigrant j from household $\imath$, for $j = 1,...,M_\imath^{*(2)}$. To simplify the notation, here we will also omit the subscript $\imath$ related to the household and thus write $\pi_j^{(2)}$.

One implication of selecting immigrants through households is that other individuals (such as newborns, initially-present individuals or longitudinal individuals) can be brought in by the supplementary sample by living in the same household as the selected immigrants. Since the selection units of the supplementary sample are restricted to the immigrants, these other individuals are indirectly selected, even if they will be surveyed. The selection probabilities of these individuals are often difficult, if not impossible, to obtain in practice.

The remaining immigrants selected for cross-sectional representativeness are those individuals who join longitudinal households, who are then considered as cohabitants. As with the newborns and initially-present individuals of the previous paragraph, the addition of these cohabitants to longitudinal households results in the inclusion of individuals having selection probabilities that are often difficult, if not impossible, to obtain in practice.

The individuals with unknown selection probabilities have entered the survey process in an indirect way. They complicate the determination of the estimation weights, as their selection probability is unknown. In order to override this difficulty, the GWSM is proposed.

6.3 USE OF THE GWSM IN OBTAINING ESTIMATION WEIGHTS

The GWSM is now applied to the SLID sample, including the supplementary sample. The population U^A is here represented by the union of the two distinct populations $U^{(1)}$ and $U^{*(2)}$, i.e., $U^A = U^* = U^{(1)} \bigcup U^{*(2)}$. The sample s^A of $m^A = m^{(1)} + m^{*(2)}$ individuals corresponds to the union of the two distinct samples $s^{(1)}$ and $s^{*(2)}$. The population U^B is represented by $U = U^{(1)} \bigcup U^{(2)}$. Note that the population $U^A = U^*$ excludes the newborns while the population $U^B = U$ includes them. The clusters of population U^B simply correspond to the N households of wave 2, and hence $M_i^B = M_i$.

A linkage between population U^A and U^B can be established by the same individuals in populations U^A and U^B. That is, $l_{j,ik}$=1 if individual *j* in population U^A corresponds to individual *k* of household *i* in population U^B, and $l_{j,ik}$=0 otherwise. Thus, these links form a one-to-one relation. For each individual *ik* not being a newborn, we then have $L_{ik}^B = \sum_{j=1}^{M^A} l_{j,ik} = 1$. On the other hand, for each newborn *ik*, we have $L_{ik}^B = \sum_{j=1}^{M^A} l_{j,ik} = 0$ since they are excluded from U^A. We now have $L_i^B = \sum_{k=1}^{M_i^B} L_{ik}^B = M_i^{*B}$ where M_i^{*B} is the size of household *i* excluding the newborns. This situation can be illustrated by Figure 6.2.

By considering definition (2.2), the initial weight w'_{ik} of individual k in household i is given by

$$w'_{ik} = \frac{t_{ik}^{(1)}}{\pi_{ik}^{(1)}} + \frac{t_{ik}^{*(2)}}{\pi_{ik}^{*(2)}}, \tag{6.1}$$

where $t_{ik}^{(1)} = 1$ if individual ik is part of $s^{(1)}$, and 0 otherwise, and $t_{ik}^{*(2)} = 1$ if individual ik is part of $s^{*(2)}$, and 0 otherwise.

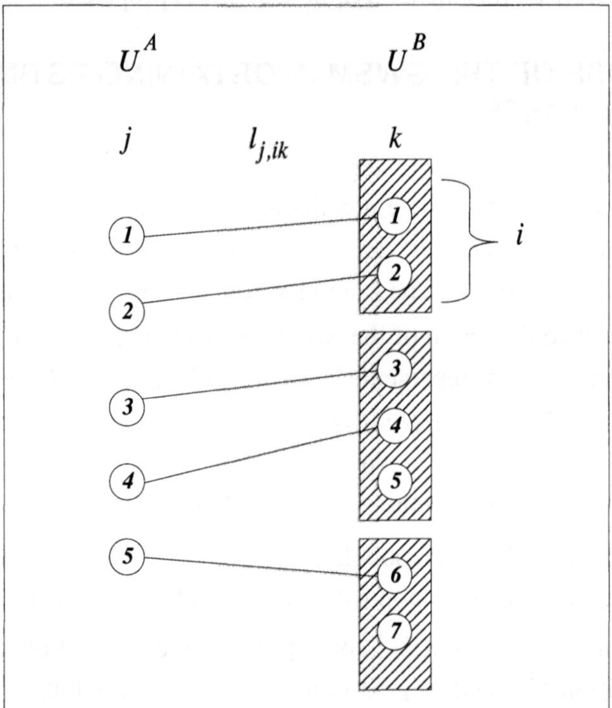

Figure 6.2: *Example of links in longitudinal surveys*

This can be written more explicitly by expressing w'_{ik} as follows:

$$w'_{ik} = \begin{cases} 1/\pi_{ik}^{(1)} & \text{for } ik \in s^{(1)} \\ 1/\pi_{ik}^{*(2)} & \text{for } ik \in s^{*(2)} \\ 0 & \text{otherwise.} \end{cases} \tag{6.2}$$

Note that the first line of (6.2) corresponds to the longitudinal individuals. The second line corresponds to the immigrants selected through the supplementary sample. The third line represents altogether newborns, cohabitants (if the household is a longitudinal household not part of the supplementary sample) and/or initially-present individuals (if the household is part of the supplementary sample).

From (2.4), the final weight w_i of household i is obtained from

$$w_i = \frac{\sum_{k=1}^{M_i^B} w'_{ik}}{\sum_{k=1}^{M_i^B} L_{ik}^B} = \frac{1}{M_i^{*B}} \sum_{k=1}^{M_i^B} w'_{ik} = \frac{1}{M_i^{*B}} \left[\sum_{k=1}^{M_i^{(1)}} \frac{t_{ik}^{(1)}}{\pi_{ik}^{(1)}} + \sum_{k=1}^{M_i^{*(2)}} \frac{t_{ik}^{*(2)}}{\pi_{ik}^{*(2)}} \right]. \quad (6.3)$$

Finally, the estimation weight w_{ik} is obtained by setting $w_{ik} = w_i$ for all individuals k of household i surveyed.

Example 6.1

As an example, take the case illustrated by Figure 6.2. Suppose that units j=1, 2, 3 are selected from U^A. Before applying the GWSM, we are going to re-index the units of U^B from Figure 6.2 according to the notation used.

Units of U^B from Fig. 6.2	1	2	3	4	5	6	7
i	1	1	2	2	2	3	3
k	1	2	1	2	3	1	2

By selecting unit j=1, we survey all units of cluster i=1. Likewise, by selecting unit j=2, we again survey the units of cluster i=1. By selecting unit j=3, we survey all units of cluster i=2. Therefore, $\Omega^B = \{1,\ 2\}$. For each unit k from clusters i of Ω^B, the initial weight w'_{ik}, the number of links L_{ik}^B and the final weight w_i are calculated:

i	k	w'_{ik}	L^B_{ik}	w_i
1	1	$\frac{1}{\pi^A_{11}}$	1	$\frac{1}{2}\left[\frac{1}{\pi^A_{11}}+\frac{1}{\pi^A_{12}}\right]$
1	2	$\frac{1}{\pi^A_{12}}$	1	$\frac{1}{2}\left[\frac{1}{\pi^A_{11}}+\frac{1}{\pi^A_{12}}\right]$
2	1	$\frac{1}{\pi^A_{21}}$	1	$\frac{1}{2}\left[\frac{1}{\pi^A_{21}}+0+0\right]=\frac{1}{2\pi^A_{21}}$
2	2	0 (because $t_{22}=0$)	1	$\frac{1}{2\pi^A_{21}}$
2	3	0 (because $l_{j,23}=0$ for all j)	0	$\frac{1}{2\pi^A_{21}}$

The estimator $\hat{Y}^B$ given by (2.1) is finally written

$$\hat{Y}^B=\frac{1}{2}\left[\frac{1}{\pi^A_{11}}+\frac{1}{\pi^A_{12}}\right]y_{11}+\frac{1}{2}\left[\frac{1}{\pi^A_{11}}+\frac{1}{\pi^A_{12}}\right]y_{12}+\frac{y_{21}}{2\pi^A_{21}}+\frac{y_{22}}{2\pi^A_{21}}+\frac{y_{23}}{2\pi^A_{21}}.$$

Using the estimation weights obtained from the GWSM, one can estimate the total $Y^B=\sum_{i=1}^{N}\sum_{k=1}^{M_i^B} y_{ik}$ of the characteristic y measured at wave 2. The estimator used is the one given by equation (2.1). From Theorem 4.1, since the links are one-to-one, $\hat{Y}^B$ can be rewritten as

$$\hat{Y}^B=\sum_{j=1}^{m^{(1)}}\frac{Z_j^*}{\pi_j^{(1)}}+\sum_{j=1}^{m^{*(2)}}\frac{Z_j^*}{\pi_j^{*(2)}}=\hat{Z}^{*(1)}+\hat{Z}^{*(2)} \tag{6.4}$$

where $Z_j^*=\overline{Y}_i^*$ for individual j of U^A linked to household i of U^B, with $\overline{Y}_i^*=\sum_{k=1}^{M_i^B} y_{ik}\big/M_i^{*B}$. Thus, estimator (6.4) is the sum of two Horvitz-Thompson estimators related to $s^{(1)}$ and $s^{*(2)}$. As shown by Corollary 4.1, this estimator is unbiased for Y^B.

6.4 VARIANCE ESTIMATION

The variance formula for $\hat{Y}^B$ is provided by equation (4.11a), or (4.11b), from Corollary 4.2. However, assuming that the two samples $s^{(1)}$ and $s^{*(2)}$ are selected independently, we can see that $Var(\hat{Y}^B) = Var(\hat{Z}^{*(1)}) + Var(\hat{Z}^{*(2)})$, where each term has the form of equation (4.11a), or (4.11b). For SLID, this assumption of independence holds if the selection of the supplementary sample is done through LFS.

Considering $\hat{Z}^{*(1)}$, we can index the individuals to reflect the fact that the $m^{(1)}$ individuals were selected at wave 1 through $n^{(1)}$ households. This gives

$$\hat{Z}^{*(1)} = \sum_{j=1}^{m^{(1)}} \frac{Z_j^*}{\pi_j^{(1)}} = \sum_{\iota=1}^{n^{(1)}} \sum_{j=1}^{M_\iota^{(1)}} \frac{Z_{\iota j}^*}{\pi_{\iota j}^{(1)}} = \sum_{\iota=1}^{n^{(1)}} \frac{1}{\pi_\iota^{(1)}} \sum_{j=1}^{M_\iota^{(1)}} Z_{\iota j}^* = \sum_{\iota=1}^{n^{(1)}} \frac{Z_\iota^{*(1)}}{\pi_\iota^{(1)}}, \tag{6.5}$$

where $Z_\iota^{(1)} = \sum_{j=1}^{M_\iota^{(1)}} Z_{\iota j}^*$ since, by selecting complete households, $\pi_{\iota j}^{(1)} = \pi_\iota^{(1)}$ for individuals j of household ι. The variance $Var(\hat{Z}^{*(1)})$ is then directly obtained as

$$Var(\hat{Z}^{*(1)}) = \sum_{\iota=1}^{N^{(1)}} \sum_{\iota'=1}^{N^{(1)}} \frac{(\pi_{\iota\iota'}^{(1)} - \pi_\iota^{(1)}\pi_{\iota'}^{(1)})}{\pi_\iota^{(1)}\pi_{\iota'}^{(1)}} Z_\iota^{*(1)} Z_{\iota'}^{*(1)}. \tag{6.6}$$

Considering $\hat{Z}^{*(2)}$, the individuals can also be indexed to reflect the fact that the $m^{*(2)}$ individuals were selected at wave 2 through $n^{*(2)}$ households. Following the same steps used for $Var(\hat{Z}^{*(1)})$, $Var(\hat{Z}^{*(2)})$ is obtained as

$$Var(\hat{Z}^{*(2)}) = \sum_{\iota=1}^{N^{*(2)}} \sum_{\iota'=1}^{N^{*(2)}} \frac{(\pi_{\iota\iota'}^{*(2)} - \pi_\iota^{*(2)}\pi_{\iota'}^{*(2)})}{\pi_\iota^{*(2)}\pi_{\iota'}^{*(2)}} Z_\iota^{*(2)} Z_{\iota'}^{*(2)}, \tag{6.7}$$

where $N^{*(2)}$ is the number of households of wave 2 containing at least one immigrant and $Z_\iota^{*(2)} = \sum_{j=1}^{M_\iota^{*(2)}} Z_{\iota j}^*$.

Recall that the quantity $M_\iota^{*(2)}$ represents the number of immigrants present in household ι.

Finally, $Var(\hat{Y}^B)$ is given by

$$Var(\hat{Z}^{*(1)}) = \sum_{l=1}^{N^{(1)}} \sum_{l'=1}^{N^{(1)}} \frac{(\pi_{ll'}^{(1)} - \pi_l^{(1)} \pi_{l'}^{(1)})}{\pi_l^{(1)} \pi_{l'}^{(1)}} Z_l^{*(1)} Z_{l'}^{*(1)}$$
$$+ \sum_{l=1}^{N^{*(2)}} \sum_{l'=1}^{N^{*(2)}} \frac{(\pi_{ll'}^{*(2)} - \pi_l^{*(2)} \pi_{l'}^{*(2)})}{\pi_l^{*(2)} \pi_{l'}^{*(2)}} Z_l^{*(2)} Z_{l'}^{*(2)}. \quad (6.8)$$

The variance (6.8) may be unbiasedly estimated using an estimator derived from (4.12) or (4.13). As SLID is in fact a subsample of LFS, the Jackknife variance estimator developed for LFS (see Singh et al., 1990) may also be used, with minor modifications. In general, the *Jackknife method* works as follows: the sample first is divided into random groups (or replicates, according to the LFS terminology). Then, the random groups c are removed in turn from the sample and a new estimate $\hat{Y}_{(c)}^B$ of the total Y^B is computed. The different estimates $\hat{Y}_{(c)}^B$ are finally compared to the original estimate $\hat{Y}^B$ to obtain an estimate of the variance $Var(\hat{Y}^B)$. For further details on the Jackknife method in general, refer to Wolter (1985) and Särndal, Swensson and Wretman (1992).

Recall that the LFS is based on a stratified multi-stage design which uses an area frame. Within each first-stage stratum h, the random groups (or replicates) correspond basically to the primary sampling units. To compute the Jackknife variance estimate for the estimation of the total Y^B, the following formula can be used:

$$\hat{V}ar^{JACK}(\hat{Y}^B) = \sum_h \frac{(C_h - 1)}{C_h} \sum_{c=1}^{C_h} (\hat{Y}_{(hc)}^B - \hat{Y}^B)^2 \quad (6.9)$$

where C_h is the number of random groups in stratum h and $\hat{Y}_{(hc)}$ is the estimate of Y^B obtained after random group c in stratum h is removed. For LFS, both estimators $\hat{Y}^B$ and $\hat{Y}_{(hc)}^B$ are post-stratified based on the integrated approach of Lemaître and Dufour (1987). SLID also uses this type of post-stratification, but this is out of the scope for the present discussion.

6.5 USE OF ANOTHER TYPE OF LINKS

In the previous sections of this chapter, a link between the populations U^A and U^B was established by the individuals that are part of the two populations. Hence, $l_{j,ik}=1$ if individual j from population U^A corresponds to individual k of household i from population U^B, and $l_{j,ik}=0$ otherwise. This is a type of link among the many possibilities.

The links described in the previous paragraph can be extended to all other members of the household, i.e., by setting $l_{j,ik}=1$ for all individuals k of a household i from U^B that belong to the same household i to which individual j (from U^A) now belongs, and 0 otherwise. In other words, $l_{j,ik}=1$ if individuals j and k belong to household i. This is illustrated in Figure 6.3.

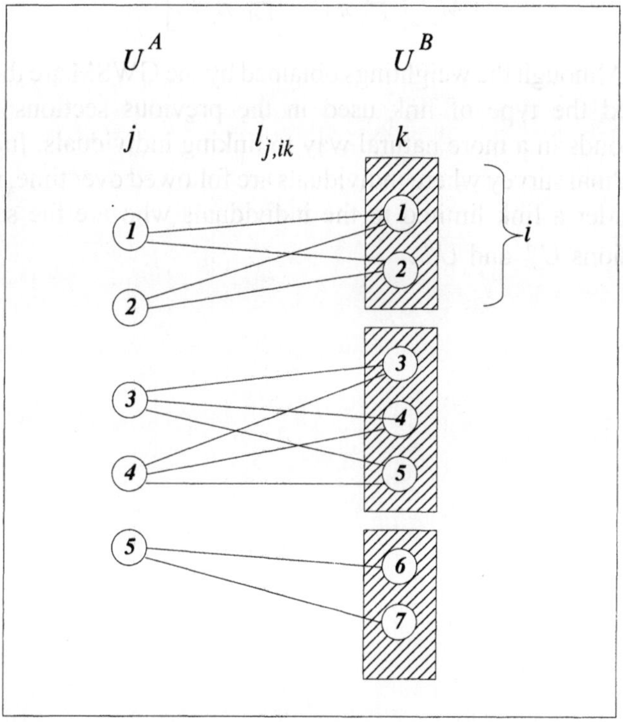

Figure 6.3: *Example of links extended to all household members*

From (2.4), the final weight is given by

$$
\begin{aligned}
w_i &= \frac{\sum_{k'=1}^{M_i^B} w'_{ik'}}{\sum_{k'=1}^{M_i^B} L_{ik'}^B} \\
&= \frac{1}{M_i^B M_i^{*B}} \sum_{k'=1}^{M_i^B} \left[\sum_{k=1}^{M_i^{(1)}} \frac{t_{ik}^{(1)}}{\pi_{ik}^{(1)}} + \sum_{k=1}^{M_i^{*(2)}} \frac{t_{ik}^{*(2)}}{\pi_{ik}^{*(2)}} \right] \\
&= \frac{1}{M_i^{*B}} \left[\sum_{k=1}^{M_i^{(1)}} \frac{t_{ik}^{(1)}}{\pi_{ik}^{(1)}} + \sum_{k=1}^{M_i^{*(2)}} \frac{t_{ik}^{*(2)}}{\pi_{ik}^{*(2)}} \right] \frac{1}{M_i^B} \sum_{k'=1}^{M_i^B} 1 \\
&= \frac{1}{M_i^{*B}} \left[\sum_{k=1}^{M_i^{(1)}} \frac{t_{ik}^{(1)}}{\pi_{ik}^{(1)}} + \sum_{k=1}^{M_i^{*(2)}} \frac{t_{ik}^{*(2)}}{\pi_{ik}^{*(2)}} \right].
\end{aligned}
\tag{6.10}
$$

Although the weightings obtained by the GWSM are the same, we preferred the type of link used in the previous sections because it corresponds in a more natural way to linking individuals. Indeed, for a longitudinal survey where individuals are followed over time, it is natural to consider a link limited to the individuals who are the same in the populations U^A and U^B.

CHAPTER 7

GWSM AND CALIBRATION

The GWSM described in the previous chapters does not use auxiliary information to obtain estimation weights. It can be imagined, however, that the use of auxiliary variables can improve the precision of estimates coming from the GWSM. For example, auxiliary information could come from the population U^A from which the sample is selected, from the target population U^B, or both of the populations. In this chapter, we are going to show that it is possible to associate the calibration of Deville and Särndal (1992) to the GWSM. In fact, it corrects the estimation weights from the GWSM so that the estimates produced correspond to known totals associated to auxiliary information. We will show that it is possible in this case to use auxiliary information from the two populations U^A and U^B. We will develop in particular the regression estimator coming from the application of calibration.

7.1 REVIEW OF CALIBRATION

Calibration arises from a generalisation by Deville (1988), and then by Deville and Särndal (1992), of an idea by Lemel (1976). Calibration consists of adjusting survey weights in such a way that the estimates are calibrated on known totals. The basic principle of calibration is to obtain estimation weights — in fact, *calibration weights* — that are the closest possible to the survey weights while satisfying the constraint that the calibrated estimates must satisfy known totals. The distance function used to measure the distance between calibration weights and survey weights determines the final form of calibration. In fact, with a judicious choice of the distance function, calibration can lead to known estimators such as the ratio estimator, the regression estimator or the raking ratio estimator.

Calibration is described as follows. From a population U of size N, a sample s of size n is selected, where each unit k is selected with probability $\pi_k > 0$. A variable of interest y_k and a column vector of auxiliary variables $\mathbf{x}_k$ of dimension p are measured, $k = 1, ..., n$. It is assumed that the total $\mathbf{X} = \sum_{k=1}^{N} \mathbf{x}_k$ is known, or at least a relatively precise estimate of this total.[1] The total Y can be estimated with the Horvitz-Thompson estimator $\hat{Y}^{HT} = \sum_{i=1}^{n} y_k / \pi_k$. However, the totals $\mathbf{X}$ are not necessarily respected in the sense where $\hat{\mathbf{X}}^{HT} = \sum_{i=1}^{n} \mathbf{x}_k / \pi_k \neq \mathbf{X}$. The problem is then to obtain calibration weights w_k^{CAL} that are the closest possible to the survey weights $d_k = 1/\pi_k$ in such a way that the totals $\mathbf{X}$ are respected, i.e., $\hat{\mathbf{X}}^{CAL} = \sum_{i=1}^{n} w_k^{CAL} \mathbf{x}_k = \mathbf{X}$. It is thus desired to minimise the changes made on the survey weights d_k.

Let $G_k(a,b)$ be a *distance function* between a and b such that:

(i) $G_k(a,b) \geq 0$;

(ii) $G_k(a,b)$ is differentiable with respect to a;

(iii) $G_k(a,b)$ is strictly convex;

(iv) $G_k(a,b)$ is defined on an interval $I_k(b)$ dependent on k and containing b;

(v) $G_k(a,a) = 0$;

(vi) $g_k(a,b) = \partial G_k(a,b) / \partial a$ is continuous and forms a one-to-one relationship between $I_k(b)$ and its image $\mathrm{Im}_k(b)$.

It then follows that $g_k(a,b)$ is strictly increasing with respect to a and that $g_k(a,a) = 0$ (Deville and Särndal, 1992).

The mathematical formulation of determining the calibration estimator $\hat{Y}^{CAL} = \sum_{k=1}^{n} w_k^{CAL} y_k$ is the following:

---

[1] Deville (2000a) discussed the problem of calibration when there is no exact value for $\mathbf{X}$, but instead an approximate value or a value estimated from another survey.

DETERMINE w_k^{CAL}, FOR $k = 1, ..., n$, TO MINIMISE

$$\sum_{k=1}^{n} G_k(w_k^{CAL}, d_k) \tag{7.1}$$

$$\text{UNDER THE CONSTRAINT } \hat{\mathbf{X}}^{CAL} = \sum_{k=1}^{n} w_k^{CAL} \mathbf{x}_k = \mathbf{X}. \tag{7.2}$$

Deville and Särndal (1992) give several examples of distance functions. In this chapter, we will restrict ourselves to the *Euclidean distance* $G_k(w_k^{CAL}, d_k) = (w_k^{CAL} - d_k)^2 / d_k$.

After having minimised the distance (7.1) under the constraint (7.2), the calibration estimator is obtained as

$$\hat{Y}^{CAL} = \sum_{k=1}^{n} w_k^{CAL} y_k = \sum_{k=1}^{n} d_k F_k(\mathbf{x}_k^T \boldsymbol{\lambda}) y_k, \tag{7.3}$$

where $w_k^{CAL} = d_k F_k(\mathbf{x}_k^T \boldsymbol{\lambda})$ is the calibration weight. Note that $F_k(\mathbf{x}_k^T \boldsymbol{\lambda})$ corresponds to the *g-weight* from Särndal, Swensson and Wretman (1992).

In the preceding formula, the function $d_k F_k(\cdot)$ is the reciprocal of $g_k(\,\cdot\,, d_k)$ that goes from $\text{Im}_k(d_k)$ to $I_k(d_k)$.

The value of the vector $\boldsymbol{\lambda}$ of dimension p is the solution of the equation $\mathbf{X} = \sum_{k=1}^{n} d_k F_k(\mathbf{x}_k^T \boldsymbol{\lambda}) \mathbf{x}_k$. Note that $\boldsymbol{\lambda}$ is the Lagrange multiplier entering the minimisation of (7.1).

To calculate the calibration weights w_k^{CAL}, Sautory (1991) and Le Guennec and Sautory (2004) developed a software program called *CALMAR*, which stands for *Calage sur marges* (or Calibration to Margins). This program produces calibration weights for the different distance functions listed by Deville and Särndal (1992). CALMAR is used in most of the surveys at the *Institut National de la Statistique et des Études Économiques* (INSEE) in France such as the *Modes de vie* (lifestyles) survey and the *Budgets de famille* (family budgets) survey. For more details, see Sautory (1992).

One can take as an example the case where the distance function is the Euclidean distance $G_k(w_k^{CAL}, d_k) = (w_k^{CAL} - d_k)^2 / d_k$. With this distance function, the *generalised regression estimator* $\hat{Y}^{REG}$ from Cassel, Särndal and Wretman (1976) is obtained. Indeed, here we obtain

$$F_k(\mathbf{x}_k^T\boldsymbol{\lambda}) = 1 + \mathbf{x}_k^T\boldsymbol{\lambda} \tag{7.4}$$

and

$$w_k^{CAL} = w_k^{REG} = d_k(1 + \mathbf{x}_k^T\boldsymbol{\lambda}) \tag{7.5}$$

with $\boldsymbol{\lambda} = (\sum_{k=1}^n d_k \mathbf{x}_k \mathbf{x}_k^T)^- (\mathbf{X} - \hat{\mathbf{X}})$. For a given square matrix **A**, the matrix $\mathbf{A}^-$ is the *generalised inverse* of **A**. Recall that the generalised inverse of **A** is any matrix $\mathbf{A}^-$ satisfying $\mathbf{A}\mathbf{A}^-\mathbf{A} = \mathbf{A}$ (Searle, 1971). If the matrix **A** is non-singular, then $\mathbf{A}^-$ is unique and furthermore $\mathbf{A}^- = \mathbf{A}^{-1}$, the inverse of **A**. The calibration estimator, with the Euclidean distance, thus has the form

$$\hat{Y}^{CAL} = \sum_{k=1}^{n} w_k^{CAL} y_k = \hat{Y}^{HT} + (\mathbf{X} - \hat{\mathbf{X}}^{HT})^T \hat{\boldsymbol{\beta}} = \hat{Y}^{REG}, \tag{7.6}$$

where $\hat{\boldsymbol{\beta}} = (\sum_{k=1}^n d_k \mathbf{x}_k \mathbf{x}_k^T)^- \sum_{k=1}^n d_k \mathbf{x}_k y_k$.

The asymptotic bias and the asymptotic variance of the calibration estimator (7.3) can be calculated. This requires first to establish an asymptotic framework. The asymptotic framework used by Deville and Särndal (1992) is essentially the same as that of Fuller and Isaki (1981), as well as Isaki and Fuller (1982). We consider a sequence of finite populations and survey designs indicated by n, the sample size.

The size N of the finite population approaches infinity with n, and we assume that for every vector **x** of variables, we have:

(i) $\lim N^{-1}\mathbf{X}$ exists;

(ii) $N^{-1}(\hat{\mathbf{X}}^{HT} - \mathbf{X}) \to \mathbf{0}$ in probability, with respect to the sampling design;

(iii) $n^{1/2} N^{-1}(\hat{\mathbf{X}}^{HT} - \mathbf{X})$ follows a multinormal $N(\mathbf{0}, \boldsymbol{\Sigma})$ distribution;

(iv) $\max_{k=1}^{n} \|\mathbf{x}_k\| = \varphi < \infty$ for all n;

(v) $\max_{k=1}^{n} \left(\left. \dfrac{\partial^2 F_k(z)}{\partial z^2} \right|_{z=0} \right) = \varphi' < \infty$ for all n.

Deville and Särndal (1992) proved that for all $\boldsymbol{F_k(\cdot)}$ satisfying the previous conditions, the calibration estimator $\hat{Y}^{CAL}$ given by (7.3) is asymptotically equivalent to the estimator $\hat{Y}^{REG}$ given by (7.6) in the sense where $N^{-1}(\hat{Y}^{CAL} - \hat{Y}^{REG}) = O_p(n^{-1})$.

This is equivalent to writing that for all $\varepsilon > 0$, there exists a whole number *n* such that $P(n \mid N^{-1}(\hat{Y}^{CAL} - \hat{Y}^{REG}) \mid < \varepsilon) = 1$. Recall that *N*, $\hat{Y}^{CAL}$, and $\hat{Y}^{REG}$ all depend on *n*. It is said that $nN^{-1}(\hat{Y}^{CAL} - \hat{Y}^{REG})$ converges toward 0 in probability.

Having given that the estimators $\hat{Y}^{CAL}$ and $\hat{Y}^{REG}$ are asymptotically equivalent, the asymptotic bias and the asymptotic variance of $\hat{Y}^{CAL}$ are the same as those for $\hat{Y}^{REG}$. The estimator $\hat{Y}^{REG}$ can be proven to be asymptotically unbiased (Särndal, Swensson and Wretman, 1992). Furthermore, the asymptotic variance of the estimator $\hat{Y}^{REG}$, and therefore of $\hat{Y}^{CAL}$, is given by

$$Var(\hat{Y}^{CAL}) \cong \sum_{k=1}^{N} \sum_{k'=1}^{N} \frac{(\pi_{kk'} - \pi_k \pi_{k'})}{\pi_k \pi_{k'}} e_k e_{k'}, \tag{7.7}$$

where $e_k = y_k - \mathbf{x}_k^T \boldsymbol{\beta}$ is the regression residual, and where the regression coefficient $\boldsymbol{\beta}$ satisfies $(\sum_{k=1}^{N} \mathbf{x}_k \mathbf{x}_k^T)\boldsymbol{\beta} = \sum_{k=1}^{N} \mathbf{x}_k y_k$. For a proof of obtaining the variance (7.7), see Särndal, Swensson and Wretman (1992).

To estimate the variance (7.7), and thus to obtain a variance estimator for $\hat{Y}^{CAL}$, Deville and Särndal (1992) suggest to use

$$\hat{V}ar(\hat{Y}^{CAL}) = \sum_{k=1}^{n} \sum_{k'=1}^{n} \frac{(\pi_{kk'} - \pi_k \pi_{k'})}{\pi_{kk'}} w_k^{CAL} \hat{e}_k^{CAL} w_{k'}^{CAL} \hat{e}_{k'}^{CAL}, \tag{7.8}$$

where $\hat{e}_k^{CAL} = y_k - \mathbf{x}_k^T \hat{\boldsymbol{\beta}}^{CAL}$ with the regression coefficient $\hat{\boldsymbol{\beta}}^{CAL}$ satisfying $(\sum_{k=1}^{n} w_k^{CAL} \mathbf{x}_k \mathbf{x}_k^T)\hat{\boldsymbol{\beta}}^{CAL} = \sum_{k=1}^{n} w_k^{CAL} \mathbf{x}_k y_k$.

The variance of $\hat{Y}^{CAL}$ can also be estimated by using the Jackknife method. In the case where the sample *s* was selected using a multi-stage design with the first stage divided into strata *h*, a variance estimator can be used that is comparable to the Jackknife estimator (6.9) described in section 6.4. This estimator is here given by

$$\hat{V}ar^{JACK}(\hat{Y}^{CAL}) = \sum_{h} \frac{(C_h - 1)}{C_h} \sum_{c=1}^{C_h} (\hat{Y}_{(hc)}^{CAL} - \hat{Y}^{CAL})^2, \tag{7.9}$$

where C_h represents the number of random groups from stratum h and $\hat{Y}_{(hc)}^{CAL}$ represents the estimate of Y obtained after the elimination of random group c in stratum h.

From the CALMAR software program, Bernier and Lavallée (1994) created the *CALJACK* software which, in addition to calculating calibration weights, performs the calculation of estimates of totals and ratios, and obtains an estimate of their variance. As the name of the software suggests, CALJACK carries out the variance estimates using the Jackknife method. CALJACK is used notably for the production of estimates in SLID (Lavallée, 1995).

Deville (1998b) developed a generalised theory for calibration. The basic idea is to generalise the function $F_k(.)$ input into the calibration estimator $\hat{Y}^{CAL}$ given by (7.3). For each unit k of the population, a calibration function $\widehat{F}_k$ is associated that goes from $\Re^p$ toward $\Re$.

The function $\widehat{F}_k$ is such that:

(i) $\widehat{F}_k(\mathbf{0}) = 1$;

(ii) $\widehat{F}_k$ is regular.

From this function, calibration equations similar to (7.2) can be solved that here take the form:

$$\sum_{k=1}^{n} d_k \widehat{F}_k(\boldsymbol{\lambda}) \mathbf{x}_k = \mathbf{X}. \tag{7.10}$$

As in the formulation given by (7.1) and (7.2), we obtain

$$\boldsymbol{\lambda} = \left(\sum\nolimits_{i=1}^{n} d_k \nabla_k \mathbf{x}_k^T\right)^{-} (\mathbf{X} - \hat{\mathbf{X}}^{HT}) + O\left(\left\|\mathbf{X} - \hat{\mathbf{X}}^{HT}\right\|^2\right), \tag{7.11}$$

where $\nabla_k = \text{grad } \widehat{F}_k(\mathbf{0})$ is a column vector of dimension p. The *generalised calibration* estimator $\hat{Y}^{CALG}$ is then given by

$$\hat{Y}^{CALG} = \sum_{k=1}^{n} w_k^{CALG} y_k = \sum_{k=1}^{n} d_k \widehat{F}_k(\boldsymbol{\lambda}) y_k. \tag{7.12}$$

The function $\widehat{F}_k$ eliminates, in a way, the explicit expression of the distance function G_k in the formulation (7.1). It thus allows a generalisation of the distance function G_k and, consequently, of calibration. The simplest particular case is the linear case where we simply take

$$\widehat{F}_k(\boldsymbol{\lambda}) = 1 + \nabla_k^T \boldsymbol{\lambda}. \tag{7.13}$$

Deville (1998b) give the variable ∇_k the name *instrumental variable*. From (7.13), by solving the calibration equations given by (7.10), the following generalised calibration estimator is then obtained:

$$\begin{aligned} \hat{Y}^{CALG} &= \hat{Y}^{HT} + (\mathbf{X} - \hat{\mathbf{X}}^{HT})^T \left(\sum\nolimits_{k=1}^{n} d_k \nabla_k \mathbf{x}_k^T\right)^{-} \sum\nolimits_{k=1}^{n} d_k \nabla_k y_k \\ &= \hat{Y}^{HT} + (\mathbf{X} - \hat{\mathbf{X}}^{HT})^T \hat{\widehat{\boldsymbol{\beta}}}. \end{aligned} \tag{7.14}$$

Note that $\hat{\widehat{\boldsymbol{\beta}}}$ is the solution of

$$\sum\nolimits_{k=1}^{n} d_k \nabla_k (y_k - \mathbf{x}_k^T \hat{\widehat{\boldsymbol{\beta}}}) = \mathbf{0}. \tag{7.15}$$

The generalised calibration weight w_k^{CALG} associated with each unit k of sample s is then given by:

$$\begin{aligned} w_k^{CALG} &= d_k + d_k \nabla_k^T \boldsymbol{\lambda} \\ &= d_k + d_k \nabla_k^T \left(\sum\nolimits_{k=1}^{n} d_k \nabla_k \mathbf{x}_k^T\right)^{-} (\mathbf{X} - \hat{\mathbf{X}}^{HT}). \end{aligned} \tag{7.16}$$

Deville (1998b) mentioned that the asymptotic variance of the generalised calibration estimator $\hat{Y}^{CALG}$ given by (7.12) is the same as the one obtained in the linear case (7.13). Thus, the asymptotic variance of the estimator $\hat{Y}^{CALG}$ is given by

$$Var(\hat{Y}^{CALG}) \cong \sum_{k=1}^{N} \sum_{k'=1}^{N} \frac{(\pi_{kk'} - \pi_k \pi_{k'})}{\pi_k \pi_{k'}} \widehat{e}_k \widehat{e}_{k'}, \tag{7.17}$$

where $\widehat{e}_k = y_k - \mathbf{x}_k^T \widehat{\boldsymbol{\beta}}$ is the regression residual and the regression coefficient $\widehat{\boldsymbol{\beta}}$ satisfies $(\sum_{k=1}^{N} \nabla_k \mathbf{x}_k^T) \widehat{\boldsymbol{\beta}} = \sum_{k=1}^{N} \nabla_k y_k$. It is important to note that the variance (7.17) depends on the instrumental variable ∇_k. Hence, two generalised calibration estimators that use different instrumental variables are not asymptotically equivalent. To estimate the variance (7.17), we can use

$$\hat{V}ar(\hat{Y}^{CALG}) = \sum_{k=1}^{n} \sum_{k'=1}^{n} \frac{(\pi_{kk'} - \pi_k \pi_{k'})}{\pi_{kk'}} w_k^{CALG} \hat{\widehat{e}}_k^{CALG} w_{k'}^{CALG} \hat{\widehat{e}}_{k'}^{CALG}, \tag{7.18}$$

where $\hat{\widehat{e}}_k^{CALG} = y_k - \mathbf{x}_k^T \hat{\widehat{\boldsymbol{\beta}}}^{CALG}$ with the regression coefficient $\hat{\widehat{\boldsymbol{\beta}}}^{CALG}$ satisfying

$$\left(\sum_{k=1}^{n} w_k^{CALG} \nabla_k \mathbf{x}_k^T\right) \hat{\hat{\boldsymbol{\beta}}}^{CALG} = \sum_{k=1}^{n} w_k^{CALG} \nabla_k y_k .$$

Another description of generalised calibration theory is found in Deville (2000b).

7.2 GWSM WITH CALIBRATION

As mentioned in the introduction, we can have auxiliary variables relating to the population U^A contained in a column vector $\mathbf{x}_j^A$ of dimension p^A for $j \in U^A$. Assume that the total $\mathbf{X}^A = \sum_{j=1}^{M^A} \mathbf{x}_j^A$ is known. We then want the estimates obtained from the auxiliary variables $\mathbf{x}_j^A$ to be equal to the known total $\mathbf{X}^A$.

In the same way, we can also have a vector of auxiliary variables relating to the target population U^B. These variables are contained in a column vector $\mathbf{x}_{ik}^B$ of dimension p^B for $ik \in U^B$. Assume that the total $\mathbf{X}^B = \sum_{i=1}^{N} \sum_{k=1}^{M_i^B} \mathbf{x}_{ik}^B$ is known. We also want the estimates obtained from the $\mathbf{x}_{ik}^B$ to be equal to the known total $\mathbf{X}^B$.

It is important to mention that it is not necessary to know the values of $\mathbf{x}_j^A$ and $\mathbf{x}_{ik}^B$ for each $j \in U^A$ and each $ik \in U^B$, but only for the units of U^A and U^B that were selected in the sample s^A or surveyed in the target population U^B.

The calibration constraints associated with the GWSM can be expressed here in the following way:

$$\hat{\mathbf{X}}^{CAL,A} = \sum_{j=1}^{m^A} w_j^{CAL,A} \mathbf{x}_j^A = \mathbf{X}^A \tag{7.19}$$

AND

$$\hat{\mathbf{X}}^{CAL,B} = \sum_{i=1}^{n} \sum_{k=1}^{M_i^B} w_{ik}^{CAL,B} \mathbf{x}_{ik}^B = \mathbf{X}^B , \tag{7.20}$$

where $w_j^{CAL,A}$ is the calibration weight obtained from the sampling weights $d_j^A = 1/\pi_j^A$. The weight $w_{ik}^{CAL,B}$ is the calibration weight of unit *k* from the surveyed cluster *i* where the GWSM was applied. This weight can be obtained by using Theorem 4.1. First of all, from (2.1), we have

$$\hat{\mathbf{X}}^B = \sum_{i=1}^{n}\sum_{k=1}^{M_i^B} w_{ik}\mathbf{x}_{ik}^B . \tag{7.21}$$

Let $\gamma_{ik} = \mathbf{X}_i^B / L_i^B$ where $\mathbf{X}_i^B = \sum_{k=1}^{M_i^B}\mathbf{x}_{ik}^B$ for all $k \in U_i^B$. From Theorem 4.1, we can then write

$$\hat{\mathbf{X}}^B = \sum_{j=1}^{M^A}\frac{t_j}{\pi_j^A}\sum_{i=1}^{N}\sum_{k=1}^{M_i^B} l_{j,ik}\gamma_{ik} = \sum_{j=1}^{M^A}\frac{t_j}{\pi_j^A}\mathbf{\Gamma}_j = \sum_{j=1}^{m^A} d_j^A\mathbf{\Gamma}_j . \tag{7.22}$$

Note that Deville (1998a) obtained a similar result using matrix notation. Since the estimator $\hat{\mathbf{X}}^B$ can also be written as a function of the units $j \in s^A$, constraint (7.20) can be rewritten under the form

$$\hat{\mathbf{X}}^{CAL,B} = \sum_{j=1}^{m^A} w_j^{CAL,A}\mathbf{\Gamma}_j = \mathbf{X}^B . \tag{7.23}$$

The two constraints (7.19) and (7.23) are now expressed as a function of the units $j \in s^A$. Let $\mathbf{x}_j^{T,AB} = (\mathbf{x}_j^{T,A}, \mathbf{\Gamma}_j^T)$ and $\mathbf{X}^{T,AB} = (\mathbf{X}^{T,A}, \mathbf{X}^{T,B})$ be column vectors of dimension $p^{AB} = p^A + p^B$. From (7.19) and (7.23), a unique constraint can then be obtained:

$$\hat{\mathbf{X}}^{CAL,AB} = \sum_{j=1}^{m^A} w_j^{CAL,A}\mathbf{x}_j^{AB} = \mathbf{X}^{AB} . \tag{7.24}$$

Recall now that with the GWSM, an estimator of the form (2.1) is developed where the weight w_{ik} of each unit k from cluster i is given by equation (2.4). By Theorem 4.1, this estimator can be rewritten as a function of units j sampled from U^A, i.e., $\hat{Y}^B = \sum_{j=1}^{M^A} t_j Z_j / \pi_j^A = \sum_{j=1}^{m^A} d_j^A Z_j$. The calibration estimator $\hat{Y}^{CAL,B}$ = $\sum_{j=1}^{m^A} w_j^{CAL,A} Z_j$ associated with the GWSM can finally be determined from the following formulation:

DETERMINE $w_j^{CAL,A}$, FOR $j = 1, \ldots, m^A$, IN ORDER TO MINIMISE

$$\sum_{j=1}^{m^A} G_j(w_j^{CAL,A}, d_j^A) \tag{7.25}$$

UNDER THE CONSTRAINT $\hat{\mathbf{X}}^{CAL,AB} = \sum_{j=1}^{m^A} w_j^{CAL,A}\mathbf{x}_j^{AB} = \mathbf{X}^{AB}$. (7.26)

This form proves to be very useful. Indeed, it corresponds exactly to the formulation of Deville and Särndal (1992) given by (7.1) and (7.2). Thus, the estimator $\hat{Y}^{CAL,B}$ can be developed to estimate the total Y^B using auxiliary variables associated to the populations U^A and U^B. After having minimised the distance (7.25) under the constraint (7.26), the calibration estimator is obtained:

$$\hat{Y}^{CAL,B} = \sum_{j=1}^{m^A} w_j^{CAL,A} Z_j = \sum_{j=1}^{m^A} d_j^A F_j(\mathbf{x}_j^{T,AB} \boldsymbol{\lambda}^{AB}) Z_j , \qquad (7.27)$$

where $w_j^{CAL,A} = d_j^A F_j(\mathbf{x}_j^{T,AB} \boldsymbol{\lambda}^{AB})$ is the calibration weight. The value of the vector $\boldsymbol{\lambda}^{AB}$ of dimension p^{AB} is the solution of the equation $\mathbf{X}^{AB} = \sum_{j=1}^{m^A} d_j^A F_j(\mathbf{x}_j^{T,AB} \boldsymbol{\lambda}^{AB}) \mathbf{x}_j^{AB}$.

As an example, we can again consider the Euclidean distance. The following calibration weight is then obtained:

$$w_j^{CAL,A} = w_j^{REG,A} = d_j^A (1 + \mathbf{x}_j^{T,AB} \boldsymbol{\lambda}^{AB}), \qquad (7.28)$$

where $\boldsymbol{\lambda}^{AB} = (\sum_{j=1}^{m^A} d_j^A \mathbf{x}_j^{AB} \mathbf{x}_j^{T,AB})^{-} (\mathbf{X}^{AB} - \hat{\mathbf{X}}^{AB})$.

The calibration (or regression) estimator given by (7.6) here takes the form:

$$\hat{Y}^{REG,B} = \sum_{j=1}^{m^A} w_j^{REG,A} Z_j = \hat{Y}^B + (\mathbf{X}^{AB} - \hat{\mathbf{X}}^{AB})^T \hat{\boldsymbol{\beta}}^{AB} , \qquad (7.29)$$

where
$$\hat{\boldsymbol{\beta}}^{AB} = \left(\sum_{j=1}^{m^A} d_j^A \mathbf{x}_j^{AB} \mathbf{x}_j^{T,AB}\right)^{-} \sum_{j=1}^{m^A} d_j^A \mathbf{x}_j^{AB} Z_j . \qquad (7.30)$$

The expression for $\hat{Y}^B$ is given by (2.1) or (4.1), and $\hat{\mathbf{X}}^{AB} = \sum_{j=1}^{m^A} d_j^A \mathbf{x}_j^{AB}$.

To obtain the calibration weight $w_{ik}^{CAL,B}$ associated with each unit k of cluster i surveyed from the target population U^B, the GWSM is applied by replacing $1/\pi_j^A$ with the calibration weight $w_j^{CAL,A}$ in (2.2).

Steps of the GWSM with calibration weights

Step 1: For each unit k of clusters i from Ω^B, calculate the initial weight $w_{ik}'^{CAL}$, to know:

$$w_{ik}'^{CAL} = \sum_{j=1}^{m^A} l_{j,ik} \, w_j^{CAL,A} . \qquad (7.31)$$

Step 2: For each unit k of clusters i from Ω^B, obtain the total number of links $L^B_{ik} = \sum_{j=1}^{M^A} l_{j,ik}$.

Step 3: Calculate the final weight $w_i^{CAL,B}$:

$$w_i^{CAL,B} = \frac{\sum_{k=1}^{M_i^B} w_{ik}^{\prime CAL}}{\sum_{k=1}^{M_i^B} L_{ik}^B}. \tag{7.32}$$

Step 4: Finally, we set $w_{ik}^{CAL,B} = w_i^{CAL,B}$ for all $k \in U_i^B$.

Following steps 1 to 4, it is concluded that

$$w_{ik}^{CAL,B} = \sum_{j=1}^{m^A} w_j^{CAL,A} \frac{L_{j,i}}{L_i^B}. \tag{7.33}$$

The estimator $\hat{Y}^{CAL,B} = \sum_{j=1}^{m^A} w_j^{CAL,A} Z_j$ determined from (7.25) and (7.26) can thus be rewritten under the form:

$$\hat{Y}^{CAL,B} = \sum_{i=1}^{n}\sum_{k=1}^{M_i^B} w_{ik}^{CAL,B} y_{ik}. \tag{7.34}$$

It is important to note that the GWSM is applied here after calibration has been performed. It will be shown in section 7.4.2 that it is possible to first apply the GWSM, and then perform calibration, in the case where auxiliary information only comes from the population U^B.

By following the proof of Theorem 4.1, it can be verified here that $\hat{\mathbf{X}}^{CAL,B} = \sum_{i=1}^{n}\sum_{k=1}^{M_i^B} w_{ik}^{CAL,B}\mathbf{x}_{ik}^B = \mathbf{X}^B$. Indeed, following Theorem 4.1, expression (7.35) is obtained.

$$\hat{\mathbf{X}}^{CAL,B} = \sum_{i=1}^{n}\sum_{k=1}^{M_i^B} w_{ik}^{CAL,B}\mathbf{x}_{ik}^B = \sum_{j=1}^{m^A} w_j^{CAL,A}\boldsymbol{\Gamma}_j. \tag{7.35}$$

Because with the calibration weight $w_j^{CAL,A}$, the constraint (7.26) is satisfied and $\mathbf{x}_j^{T,AB} = (\mathbf{x}_j^{T,A}, \boldsymbol{\Gamma}_j^T)$ and $\mathbf{X}^{T,AB} = (\mathbf{X}^{T,A}, \mathbf{X}^{T,B})$, we directly get

$$\sum_{j=1}^{m^A} w_j^{CAL,A}\boldsymbol{\Gamma}_j = \mathbf{X}^B. \tag{7.36}$$

As the estimator $\hat{Y}^{CAL,B}$ is obtained by the formulations (7.25) and (7.26) that correspond exactly to the formulation from Deville and Särndal (1992), the asymptotic bias and the asymptotic variance of $\hat{Y}^{CAL,B}$ are then the same as those for $\hat{Y}^{REG,B}$. For these asymptotic properties, we consider here a sequence of finite populations U^A and sampling designs indexed by the sample size m^A. The size M^A of finite population U^A approaches infinity with m^A.

The estimator $\hat{Y}^{REG,B}$ given by (7.29) is nothing more than the estimator $\hat{Y}^{REG}$ given by (7.6) where the variable y_k is replaced by the variable Z_j. Thus, the estimator $\hat{Y}^{CAL,B}$ is asymptotically unbiased. Furthermore, the asymptotic variance of the estimator $\hat{Y}^{CAL,B}$ is given by

$$Var(\hat{Y}^{CAL,B}) \cong \sum_{j=1}^{M^A}\sum_{j'=1}^{M^A}\frac{(\pi_{jj'}^A - \pi_j^A\pi_{j'}^A)}{\pi_j^A\pi_{j'}^A}e_j^A e_{j'}^A, \qquad (7.37)$$

where $e_j^A = Z_j - \mathbf{x}_j^{T,AB}\boldsymbol{\beta}^{AB}$ is the regression residual and where the regression coefficient $\boldsymbol{\beta}^{AB}$ satisfies $(\sum_{j=1}^{M^A}\mathbf{x}_j^{AB}\mathbf{x}_j^{T,AB})\boldsymbol{\beta}^{AB} = \sum_{j=1}^{M^A}\mathbf{x}_j^{AB}Z_j$. To estimate the variance (7.37), we can follow the suggestion from Deville and Särndal (1992) and use

$$\hat{V}ar(\hat{Y}^{CAL,B}) = \sum_{j=1}^{m^A}\sum_{j'=1}^{m^A}\frac{(\pi_{jj'}^A - \pi_j^A\pi_{j'}^A)}{\pi_{jj'}^A}w_j^{CAL,A}\hat{e}_j^{CAL,A}w_{j'}^{CAL,A}\hat{e}_{j'}^{CAL,A}, \qquad (7.38)$$

where $\hat{e}_j^{CAL,A} = Z_j - \mathbf{x}_j^{T,AB}\hat{\boldsymbol{\beta}}^{CAL,AB}$ with the regression coefficient $\hat{\boldsymbol{\beta}}^{CAL,AB}$ satisfying $(\sum_{j=1}^{m^A}w_j^{CAL,A}\mathbf{x}_j^{AB}\mathbf{x}_j^{T,AB})\hat{\boldsymbol{\beta}}^{CAL,AB} = \sum_{j=1}^{m^A}w_j^{CAL,A}\mathbf{x}_j^{AB}Z_j$.

The variance of $\hat{Y}^{CAL,B}$ can also be estimated by using the Jackknife method with an estimator similar to (7.9).

7.3 PARTICULAR CASE 1: AUXILIARY VARIABLES COMING FROM U^A

In the previous section, calibration was associated to the GWSM by developing the general case where auxiliary information comes from the population U^A from which the sample is selected, the target population U^B, or both of the populations. From the developed theory, the results are here derived for the particular case where auxiliary information comes from U^A only.

We have auxiliary variables relating to the population U^A and contained in a column vector $\mathbf{x}_j^A$ of dimension p^A for $j \in U^A$. Assume that the total $\mathbf{X}^A = \sum_{j=1}^{M^A} \mathbf{x}_j^A$ is known. The calibration estimator $\hat{Y}^{CAL,B} = \sum_{j=1}^{m^A} w_j^{CAL,A} Z_j$ associated with the GWSM when there are auxiliary variables $\mathbf{x}_j^A$ can be expressed in the following way:

DETERMINE $w_j^{CAL,A}$, FOR $j = 1, ..., m^A$, IN ORDER TO MINIMISE

$$\sum_{j=1}^{m^A} G_j(w_j^{CAL,A}, d_j^A) \qquad (7.39)$$

UNDER THE CONSTRAINT $\hat{\mathbf{X}}^{CAL,A} = \sum_{j=1}^{m^A} w_j^{CAL,A} \mathbf{x}_j^A = \mathbf{X}^A$. (7.40)

After having minimised the distance (7.39) under the constraint (7.40), the calibration estimator is obtained:

$$\hat{Y}^{CAL,B} = \sum_{j=1}^{m^A} w_j^{CAL,A} Z_j = \sum_{j=1}^{m^A} d_j^A F_j(\mathbf{x}_j^{T,A} \boldsymbol{\lambda}^A) Z_j \,, \qquad (7.41)$$

where $w_j^{CAL,A} = d_j^A F_j(\mathbf{x}_j^{T,A} \boldsymbol{\lambda}^A)$ is the calibration weight. The value of the vector $\boldsymbol{\lambda}^A$ of dimension p^A is the solution of $\mathbf{X}^A = \sum_{j=1}^{m^A} d_j^A F_j(\mathbf{x}_j^{T,A} \boldsymbol{\lambda}^A) \mathbf{x}_j^A$.

As an example, we can again consider the Euclidean distance. The calibration (or regression) estimator given by (7.29) and (7.30) here takes the form:

$$\hat{Y}^{REG,B} = \sum_{j=1}^{m^A} w_j^{REG,A} Z_j = \hat{Y}^B + (\mathbf{X}^A - \hat{\mathbf{X}}^A)^T \hat{\boldsymbol{\beta}}^A \qquad (7.42)$$

where
$$\hat{\boldsymbol{\beta}}^A = \left(\sum_{j=1}^{m^A} d_j^A \mathbf{x}_j^A \mathbf{x}_j^{T,A} \right)^{-} \sum_{j=1}^{m^A} d_j^A \mathbf{x}_j^A Z_j \,. \qquad (7.43)$$

The expression for $\hat{Y}^B$ is given by (2.1) or even (4.1), and $\hat{\mathbf{X}}^A = \sum_{j=1}^{m^A} d_j^A \mathbf{x}_j^A$.

To obtain the calibration weight $w_{ik}^{CAL,B}$ associated to each unit k of cluster i surveyed from the target population U^B, the GWSM is applied by replacing $1/\pi_j^A$ with the calibration weight $w_j^{CAL,A}$ in (2.2). Thus, steps 1 to 4 are performed as described in section 7.2.

In practice, the auxiliary information of U^A often comes down to qualitative variables that were not used in the stratification, or else quantitative variables on which we wish to calibrate the estimates.

Note that the theory presented is general and thus the auxiliary variables can be qualitative, quantitative, or a mix of the two. Take the example of selecting parents to survey their children (Figure 1.2). Suppose that the selection of parents was carried out from an area sampling frame. It can then prove useful to calibrate the weight of the parents on the age-sex categories (qualitative variables), and also on the income of the persons (quantitative variable). Note that stratification by age-sex groups is notably used by the LFS (Singh *et al.*, 1990, and Dufour *et al.*, 1998).

It is important to remember that the choice of calibration variables is linked to the availability of the totals $\mathbf{X}^A = \sum_{j=1}^{M^A} \mathbf{x}_j^A$. Going back to the previous example, it is clear that it is pointless to consider using income as the calibration variable if the total income or a relatively precise estimate of it is not known.

In some cases, it turns out to be useful to choose, as auxiliary variables coming from U^A, the stratification variables used for the selection of sample s^A. This is particularly the case for sampling designs with random sample sizes such as Poisson sampling. With this type of sampling where the final sample size m^A is random, it is generally noticed that $\sum_{j=1}^{m^A} d_j^A \neq M^A$. If a Horvitz-Thompson estimator is used to produce estimates, as is the case with the GWSM without calibration, the estimates produced then have very large variances (Särndal, Swensson and Wretman, 1992). To correct this problem, it is strongly advised to perform calibration using the stratification variables as auxiliary variables. For more details, see Lavallée (1998b).

A choice of an auxiliary variable that can prove to be very efficient for improving the precision of the estimates drawn from the GWSM is the number of links L_j^A. Indeed in section 5.2, it was noted, in the extreme case where the population U^B only has a single cluster, that the variance of $\hat{Y}^B$ is non-zero, though the population U^B then undergoes a census. This observation was also made in section 4.4. Part of the variance in fact comes from the complex links. By setting $\mathbf{x}_j^A = L_j^A$ for $j \in U^A$, this variance is reduced to zero by calibrating the estimator $\hat{Y}^B$ on the total number of links L. It is assumed here, of course, that the total number of links L is known or, if not, that we have a good estimate of L.

We can take, for example, the regression estimator given by (7.42). By setting $\mathbf{x}_j^A = L_j^A$, the following estimator calibrated on the total number of links L is obtained:

$$\begin{aligned}\hat{Y}^{REG,B} &= \sum_{j=1}^{m^A} w_j^{REG,A} Z_j \\ &= \hat{Y}^B + (L - \hat{L})\hat{\beta}^A\end{aligned} \tag{7.44}$$

where $$\hat{\beta}^A = \left(\sum_{j=1}^{m^A} d_j^A (L_j^A)^2\right)^{-1} \sum_{j=1}^{m^A} d_j^A L_j^A Z_j . \tag{7.45}$$

By replacing $\hat{Y}^B$ with (4.1) in (7.44), we then get

$$\begin{aligned}\hat{Y}^{REG,B} &= \sum_{j=1}^{m^A} \frac{Z_j}{\pi_j^A} + (L - \hat{L})\hat{\beta}^A \\ &= \sum_{j=1}^{m^A} \frac{Z_j}{\pi_j^A} + (L - \hat{L}) \frac{\sum_{j=1}^{m^A} d_j^A L_j^A Z_j}{\sum_{j=1}^{m^A} d_j^A (L_j^A)^2}.\end{aligned} \tag{7.46}$$

Looking at the estimator given by (7.46), we see that if the variable of interest Z_j (derived from y) is replaced by the auxiliary variable L_j^A, the resulting estimator $\hat{L}^{REG,B}$ is equal to L. Now, suppose that the population U^B only has a single cluster of size M^B, which implies that $Z_j = L_j^A \; Y^B / L$ by (5.24). In this case, the estimator (7.46) is written

$$\hat{Y}^{REG,B} = \frac{Y^B}{L}\left[\sum_{j=1}^{m^A} \frac{L_j^A}{\pi_j^A} + (L - \hat{L}) \frac{\sum_{j=1}^{m^A} d_j^A L_j^A L_j^A}{\sum_{j=1}^{m^A} d_j^A (L_j^A)^2}\right] = Y^B . \tag{7.47}$$

Thus, with a single cluster for the target population U^B, the estimator $\hat{Y}^{REG,B}$ has a zero variance. As noticed in section 5.2, by choosing only a single unit j from U^A, it should have been possible to estimate the total Y^B with a zero variance due to the fact that having only a single cluster causes a census of the population U^B. The variance of $\hat{Y}^B$ is non-zero because of the complex links that can exist between populations U^A and U^B. Fortunately, we see that the estimator (7.44) obtained by calibration corrects this situation.

7.4 PARTICULAR CASE 2: AUXILIARY VARIABLES COMING FROM U^B

In section 7.2, calibration was associated to the GWSM by developing the general case where auxiliary information comes from U^A, U^B, or both. From the theory developed, the results here are derived for the particular case where auxiliary information comes only from U^B. Note that the theory developed in section 7.2 was obtained by performing calibration **before** using the GWSM to obtain the weights $w_{ik}^{CAL,B}$ associated with the units k of surveyed clusters i. In the current section, weights $\dot{w}_{ik}^{CAL,B}$ will also be obtained that are calculated by performing calibration **after** the use of the GWSM. These two sets of weights will then be compared in order to determine which of the two is preferable.

We have auxiliary variables relating to the target population U^B. These variables are contained in a column vector $\mathbf{x}_{ik}^B$ of dimension p^B for $ik \in U^B$. Assume that the total $\mathbf{X}^B = \sum_{i=1}^{N} \sum_{k=1}^{M_i^B} \mathbf{x}_{ik}^B$ is known, or at least that there is a relatively precise estimate of this total.

7.4.1 Application of calibration before GWSM

The calibration constraint associated with the GWSM is formulated here by (7.20). This constraint was seen to be equivalent to that given by (7.23) expressed as a function of units $j \in s^A$. The calibration estimator $\hat{Y}^{CAL,B} = \sum_{j=1}^{m^A} w_j^{CAL,A} Z_j$ associated with the GWSM is determined from the following formulation:

DETERMINE $w_j^{CAL,A}$, FOR $j = 1, ..., m^A$, IN ORDER TO MINIMISE

$$\sum_{j=1}^{m^A} G_j(w_j^{CAL,A}, d_j^A) \tag{7.48}$$

UNDER THE CONSTRAINT $\hat{\mathbf{X}}^{CAL,B} = \sum_{j=1}^{m^A} w_j^{CAL,A} \mathbf{\Gamma}_j = \mathbf{X}^B$. (7.49)

After having minimised the distance (7.48) under the constraint (7.49), the calibration estimator is obtained by (7.50).

$$\hat{Y}^{CAL,B} = \sum_{j=1}^{m^A} w_j^{CAL,A} Z_j = \sum_{j=1}^{m^A} d_j^A F_j(\mathbf{\Gamma}_j^T \boldsymbol{\lambda}^B) Z_j , \tag{7.50}$$

where $w_j^{CAL,A} = d_j^A F_j(\mathbf{\Gamma}_j^T \boldsymbol{\lambda}^B)$ is the calibration weight. The value of the vector $\boldsymbol{\lambda}^B$ of dimension p^B is the solution of $\mathbf{X}^B = \sum_{j=1}^{m^A} d_j^A F_j(\mathbf{\Gamma}_j^T \boldsymbol{\lambda}^B) \mathbf{\Gamma}_j$.

As an example, consider the Euclidean distance. The calibration (or regression) estimator given by (7.29) is here:

$$\begin{aligned} \hat{Y}^{REG,B} &= \sum_{j=1}^{m^A} w_j^{REG,A} Z_j \\ &= \hat{Y}^B + (\mathbf{X}^B - \hat{\mathbf{X}}^B)^T \hat{\boldsymbol{\beta}}^B \end{aligned} \tag{7.51}$$

where $$\hat{\boldsymbol{\beta}}^B = \left(\sum_{j=1}^{m^A} d_j^A \mathbf{\Gamma}_j \mathbf{\Gamma}_j^T \right)^{-} \sum_{j=1}^{m^A} d_j^A \mathbf{\Gamma}_j Z_j . \tag{7.52}$$

The expression for $\hat{Y}^B$ is given by (2.1) or also (4.1), and $\hat{\mathbf{X}}^B = \sum_{j=1}^{m^A} d_j^A \mathbf{\Gamma}_j$.

To obtain the calibration weight $w_{ik}^{CAL,B}$ associated with each unit k of cluster i surveyed from the target population U^B, the GWSM is applied from the calibration weights $w_j^{CAL,A}$ obtained earlier. The GWSM is performed according to steps 1 to 4 described in section 7.2. The estimator $\hat{Y}^{CAL,B} = \sum_{j=1}^{m^A} w_j^{CAL,A} Z_j$ is then rewritten under the form $\hat{Y}^{CAL,B} = \sum_{i=1}^{n} \sum_{k=1}^{M_i^B} w_{ik}^{CAL,B} y_{ik}$.

As $w_{ik}^{CAL,B} = w_i^{CAL,B}$, we also have

$$\hat{Y}^{CAL,B} = \sum_{i=1}^{n} w_i^{CAL,B} Y_i .$$

As an example, with the Euclidean distance, from (7.51) and (7.52), the following calibration weight is obtained by using (7.33):

$$
\begin{aligned}
w_i^{REG,B} &= \sum_{j=1}^{m^A} w_j^{REG,A} \frac{L_{j,i}}{L_i^B} \\
&= \sum_{j=1}^{m^A} \left[d_j^A + d_j^A \mathbf{\Gamma}_j^T \left(\sum_{j=1}^{m^A} d_j^A \mathbf{\Gamma}_j \mathbf{\Gamma}_j^T \right)^{-} (\mathbf{X}^B - \hat{\mathbf{X}}^B) \right] \frac{L_{j,i}}{L_i^B} \\
&= \sum_{j=1}^{m^A} d_j^A \frac{L_{j,i}}{L_i^B} \qquad (7.53) \\
&+ \sum_{j=1}^{m^A} d_j^A \frac{L_{j,i}}{L_i^B} \left[\mathbf{\Gamma}_j^T \left(\sum_{j=1}^{m^A} d_j^A \mathbf{\Gamma}_j \mathbf{\Gamma}_j^T \right)^{-} (\mathbf{X}^B - \hat{\mathbf{X}}^B) \right] \\
&= w_i + \sum_{j=1}^{m^A} d_j^A \frac{L_{j,i}}{L_i^B} \mathbf{\Gamma}_j^T \left(\sum_{j=1}^{m^A} d_j^A \mathbf{\Gamma}_j \mathbf{\Gamma}_j^T \right)^{-} (\mathbf{X}^B - \hat{\mathbf{X}}^B).
\end{aligned}
$$

The last line follows from Result 2.1.

7.4.2 Application of calibration after GWSM

To estimate the total Y^B of the target population U^B, we have the estimator $\hat{Y}^B$ given by (2.1) and obtained from the GWSM. If we have auxiliary variables $\mathbf{x}_{ik}^B$ for which the total $\mathbf{X}^B = \sum_{i=1}^{N} \sum_{k=1}^{M_i^B} \mathbf{x}_{ik}^B$ is known, the possibility of directly calibrating the estimator $\hat{Y}^B$ on the total $\mathbf{X}^B$ can be considered. Note that this approach corresponds to that used for the calibration of estimates produced by SLID (Lavallée and Hunter, 1993, and Lévesque and Franklin, 2000).

From the weights w_{ik} given by (2.4), the calibration estimator $\hat{Y}^{CAL,B} = \sum_{i=1}^{n} \sum_{k=1}^{M_i^B} \dot{w}_{ik}^{CAL,B} y_{ik}$ (obtained after using the GWSM) can be determined from the following formulation:

DETERMINE $\dot{w}_{ik}^{CAL,B}$, FOR $k = 1,...,M_i^B$ AND $i = 1,...,n$, IN ORDER TO MINIMISE

$$\sum_{i=1}^{n} \sum_{k=1}^{M_i^B} G_{ik}(\dot{w}_{ik}^{CAL,B}, w_{ik}) \qquad (7.54)$$

UNDER THE CONSTRAINT $\hat{\mathbf{X}}^{CAL,B} = \sum_{i=1}^{n} \sum_{k=1}^{M_i^B} \dot{w}_{ik}^{CAL,B} \mathbf{x}_{ik}^B = \mathbf{X}^B$. (7.55)

Although the weight w_{ik} is not strictly a sampling weight (i.e., the inverse of the selection probability), the theory of Deville and Särndal (1992) presented in section 7.1 remains valid for the determination of the calibration weights $\dot{w}_{ik}^{CAL,B}$. After having minimised the distance (7.54) under the constraint (7.55), the calibration estimator is obtained:

$$\hat{Y}^{CAL,B} = \sum_{i=1}^{n}\sum_{k=1}^{M_i^B} \dot{w}_{ik}^{CAL,B} y_{ik} = \sum_{i=1}^{n}\sum_{k=1}^{M_i^B} w_{ik} F_{ik}(\mathbf{x}_{ik}^{T,B} \dot{\boldsymbol{\lambda}}^B) y_{ik}, \quad (7.56)$$

where $\dot{w}_{ik}^{CAL,B} = w_{ik} F_{ik}(\mathbf{x}_{ik}^{T,B} \dot{\boldsymbol{\lambda}}^B)$ is the calibration weight obtained after having used the GWSM. The value of the vector $\dot{\boldsymbol{\lambda}}^B$ of dimension p^B is the solution of $\mathbf{X}^B = \sum_{i=1}^{n}\sum_{k=1}^{M_i^B} w_{ik} F_{ik}(\mathbf{x}_{ik}^{T,B} \dot{\boldsymbol{\lambda}}^B)\mathbf{x}_{ik}^B$.

We can again take as an example the case where the distance function selected is the Euclidean distance $G_{ik}(\dot{w}_{ik}^{CAL,B}, w_{ik}) = (\dot{w}_{ik}^{CAL,B} - w_{ik})^2 / w_{ik}$. With this distance function, we get

$$F_{ik}(\mathbf{x}_{ik}^{T,B} \dot{\boldsymbol{\lambda}}^B) = 1 + \mathbf{x}_{ik}^{T,B} \dot{\boldsymbol{\lambda}}^B \quad (7.57)$$

and

$$\dot{w}_{ik}^{CAL,B} = \dot{w}_{ik}^{REG,B} = w_{ik}(1 + \mathbf{x}_{ik}^{T,B} \dot{\boldsymbol{\lambda}}^B) \quad (7.58)$$

with $\dot{\boldsymbol{\lambda}}^B = \left(\sum_{i=1}^{n}\sum_{k=1}^{M_i^B} w_{ik}\mathbf{x}_{ik}^B \mathbf{x}_{ik}^{T,B}\right)^{-} (\mathbf{X}^B - \hat{\mathbf{X}}^B)$. The calibration (or regression) estimator obtained with the Euclidean distance thus has the form

$$\hat{Y}^{CAL,B} = \hat{Y}^{REG,B} = \sum_{i=1}^{n}\sum_{k=1}^{M_i^B} \dot{w}_{ik}^{REG,B} y_{ik} = \hat{Y}^B + (\mathbf{X}^B - \hat{\mathbf{X}}^B)^T \hat{\boldsymbol{\beta}}^B, \quad (7.59)$$

where $\hat{\boldsymbol{\beta}}^B = \left(\sum_{i=1}^{n}\sum_{k=1}^{M_i^B} w_{ik}\mathbf{x}_{ik}^B \mathbf{x}_{ik}^{T,B}\right)^{-} \sum_{i=1}^{n}\sum_{k=1}^{M_i^B} w_{ik}\mathbf{x}_{ik}^B y_{ik}$. The expression for $\hat{Y}^B$ is given by (2.1) or (4.1), and $\hat{\mathbf{X}}^B$ by (7.21) or (7.22).

The asymptotic bias and the asymptotic variance of $\hat{Y}^{CAL,B}$ can be obtained by specifying the asymptotic framework for the identification of the n clusters of Ω^B. In section 7.3, a sequence of finite populations U^A and survey designs indexed by the sample size m^A were considered where the population size M^A approaches infinity with m^A. Here, a sequence of target populations U^B and a sequence of sets of clusters Ω^B are added, both indexed by m^A. The sizes M^B and N approach infinity with m^A. This addition to the asymptotic framework is natural in the context of an indirect sampling of the target population U^B, through the population U^A. Indeed, if the population U^A increases, it is natural to imagine that the population U^B can increase also, given that the two populations are linked to one another. For example, in the case of the survey of children identified from a list of parents, we can conceive that the number of children increases as quickly as the list of parents increases. The same considerations are applied for s^A and Ω^B.

With this asymptotic framework, we can go back to the results of Deville and Särndal (1992). It is obtained that $\hat{\hat{Y}}^{CAL,B}$ and $\hat{\hat{Y}}^{REG,B}$ are asymptotically equivalent. Consequently, the asymptotic bias and the asymptotic variance of $\hat{\hat{Y}}^{CAL,B}$ are the same as those for $\hat{\hat{Y}}^{REG,B}$.

To get the asymptotic bias and the asymptotic variance of $\hat{\hat{Y}}^{REG,B}$, first the estimator $\hat{\hat{Y}}^{REG,B}$ given by (7.59) is expressed as a function of units j from s^A, instead of the units k of surveyed clusters i. The expressions of $\hat{Y}^B$ and $\hat{\mathbf{X}}^B$ are already expressed in these terms by (4.1) and (7.22), respectively. It remains to rewrite the estimated parameter $\hat{\hat{\boldsymbol{\beta}}}^B$.

At the start, using Result 2.1, we have

$$\begin{aligned}\sum_{i=1}^{n}\sum_{k=1}^{M_i^B} w_{ik}\mathbf{x}_{ik}^B\mathbf{x}_{ik}^{T,B} &= \sum_{i=1}^{n}\sum_{k=1}^{M_i^B}\left(\sum_{j=1}^{M^A}\frac{t_j}{\pi_j^A}\frac{L_{j,i}}{L_i^B}\right)\mathbf{x}_{ik}^B\mathbf{x}_{ik}^{T,B}\\ &= \sum_{i=1}^{n}\left(\sum_{j=1}^{M^A}\frac{t_j}{\pi_j^A}\frac{L_{j,i}}{L_i^B}\right)\sum_{k=1}^{M_i^B}\mathbf{x}_{ik}^B\mathbf{x}_{ik}^{T,B}\\ &= \sum_{j=1}^{M^A}\frac{t_j}{\pi_j^A}\sum_{i=1}^{n}\frac{L_{j,i}}{L_i^B}\sum_{k=1}^{M_i^B}\mathbf{x}_{ik}^B\mathbf{x}_{ik}^{T,B}\\ &= \sum_{j=1}^{M^A}\frac{t_j}{\pi_j^A}\sum_{i=1}^{N}\frac{L_{j,i}}{L_i^B}\sum_{k=1}^{M_i^B}\mathbf{x}_{ik}^B\mathbf{x}_{ik}^{T,B}.\end{aligned} \qquad (7.60)$$

The last line follows from the proof of Theorem 4.1. Following the same proof as that for (7.60), the following result is obtained:

$$\sum_{i=1}^{n}\sum_{k=1}^{M_i^B} w_{ik}\mathbf{x}_{ik}^B y_{ik} = \sum_{j=1}^{M^A}\frac{t_j}{\pi_j^A}\sum_{i=1}^{N}\frac{L_{j,i}}{L_i^B}\sum_{k=1}^{M_i^B}\mathbf{x}_{ik}^B y_{ik} \,. \tag{7.61}$$

Let $\mathbf{\Psi}_j^{\mathbf{xx}} = \sum_{i=1}^{N}(L_{j,i}/L_i^B)\sum_{k=1}^{M_i^B}\mathbf{x}_{ik}^B\mathbf{x}_{ik}^{T,B}$ be the matrix of dimension $p^B \times p^B$ and $\mathbf{\psi}_j^{\mathbf{x}y} = \sum_{i=1}^{N}(L_{j,i}/L_i^B)\sum_{k=1}^{M_i^B}\mathbf{x}_{ik}^B y_{ik}$ be the column vector of dimension p^B. By replacing these expressions in (7.60) and (7.61), we finally get

$$\hat{\dot{\mathbf{\beta}}}^B = \left(\sum_{j=1}^{M^A}\frac{t_j}{\pi_j^A}\mathbf{\Psi}_j^{\mathbf{xx}}\right)^{-}\sum_{j=1}^{M^A}\frac{t_j}{\pi_j^A}\mathbf{\psi}_j^{\mathbf{x}y} \,. \tag{7.62}$$

With (7.62), (4.1) and (7.22), the estimator $\hat{Y}^{REG,B}$ given by (7.59) can finally be rewritten as a function of units j from s^A. To obtain the asymptotic bias and the asymptotic variance of $\hat{Y}^{REG,B}$, the Taylor linearisation method is applied, as suggested by Särndal, Swensson and Wretman (1992). We then obtain

$$\begin{aligned}\hat{Y}^{REG,B} &\cong \hat{Y}^B + (\mathbf{X}^B - \hat{\mathbf{X}}^B)^T\dot{\mathbf{\beta}}^B \\ &= \hat{Y}^B - \hat{\mathbf{X}}^{T,B}\dot{\mathbf{\beta}}^B + \mathbf{X}^{T,B}\dot{\mathbf{\beta}}^B = \sum_{j=1}^{M^A}\frac{t_j}{\pi_j^A}(Z_j - \mathbf{\Gamma}_j^T\dot{\mathbf{\beta}}^B) + \mathbf{X}^{T,B}\dot{\mathbf{\beta}}^B\end{aligned} \tag{7.63}$$

where the regression coefficient $\dot{\mathbf{\beta}}^B$ satisfies $(\sum_{j=1}^{M^A}\mathbf{\Psi}_j^{\mathbf{xx}})\dot{\mathbf{\beta}}^B = \sum_{j=1}^{M^A}\mathbf{\psi}_j^{\mathbf{x}y}$.

As $E(\hat{Y}^{REG,B}) \cong Z = Y^B$, we see that this estimator is asymptotically unbiased. Furthermore, its asymptotic variance is given by

$$Var(\hat{Y}^{REG,B}) \cong \sum_{j=1}^{M^A}\sum_{j'=1}^{M^A}\frac{(\pi_{jj'}^A - \pi_j^A\pi_{j'}^A)}{\pi_j^A\pi_{j'}^A}\dot{e}_j^A\dot{e}_{j'}^A \tag{7.64}$$

where $\dot{e}_j^A = Z_j - \mathbf{\Gamma}_j^T\dot{\mathbf{\beta}}^B$ is the regression residual.

To estimate the variance (7.64), we can use

$$\hat{V}ar(\hat{Y}^{REG,B}) = \sum_{j=1}^{m^A}\sum_{j'=1}^{m^A}\frac{(\pi_{jj'}^A - \pi_j^A\pi_{j'}^A)}{\pi_{jj'}^A\pi_j^A\pi_{j'}^A}\hat{\dot{e}}_j^A\hat{\dot{e}}_{j'}^A \tag{7.65}$$

where $\hat{e}_j^A = Z_j - \boldsymbol{\Gamma}_j^T \hat{\hat{\boldsymbol{\beta}}}^B$. It is also possible to estimate the variance of $\hat{\hat{Y}}^{REG,B}$ using the Jackknife method with an estimator similar to (7.9).

7.4.3 Comparison of the two approaches

To estimate the total Y^B, we now have two estimators where calibration has been performed. It is a matter of estimators $\hat{Y}^{CAL,B}$ and $\hat{\dot{Y}}^{CAL,B}$ given respectively by (7.50) and (7.56). Recall that the first is obtained by performing calibration before using the GWSM, while the second is obtained by performing calibration after using the GWSM. These estimators are different, although the two are asymptotically unbiased, and though for the two sets of weights $w_{ik}^{CAL,B}$ and $\dot{w}_{ik}^{CAL,B}$, we have, respectively, $\hat{\mathbf{X}}^{CAL,B} = \mathbf{X}^B$ and $\hat{\dot{\mathbf{X}}}^{CAL,B} = \mathbf{X}^B$.

The main difference between the two estimators lays in the fact that $\hat{Y}^{CAL,B}$ has weights $w_{ik}^{CAL,B}$ that are identical for all units k of the clusters i from Ω^B, which is not the case for $\hat{\dot{Y}}^{CAL,B}$. This is quite clear if we consider step 4 of the GWSM that assigns to units k of cluster i the weight $w_i^{CAL,B}$. On the other hand, for $\hat{\dot{Y}}^{CAL,B}$, although the weights w_{ik} before calibration are identical for all units k of the clusters i from Ω^B, nothing guarantees that the weights $\dot{w}_{ik}^{CAL,B}$ after calibration are always identical. This depends on the choice of the auxiliary variables $\mathbf{x}_{ik}^B$. Many examples are found in practice where the weights $\dot{w}_{ik}^{CAL,B}$ are different within the same cluster after calibration. The most common example is that of household surveys, such as LFS, where the estimates are calibrated according to age-sex groups (Singh *et al.*, 1990, and Dufour *et al.*, 1998). Since in general a household has individuals belonging to different age-sex groups, different calibration weights are in practice obtained within a household, even though the weights before calibration are identical.

It is possible to force the calibration weights to be identical within a household by using *integrated weighting*. A method of integrated weighting described by Sautory (1993) is notably used by the *Institut National de la Statistique et des Études Économiques* (INSEE) in France. This method consists of considering the sampling unit as being the cluster, instead of the unit itself. We then no longer work with the units *ik* of the target population U^B, but rather with the clusters *i*. Note that the integrated weighting method at INSEE differs from that used by Statistics Canada. To learn more on this last method, see Lemaître and Dufour (1987).

With integrated weighting, it is assumed that we have auxiliary variables $\mathbf{X}_i^B = \sum_{k=1}^{M_i^B} \mathbf{x}_{ik}^B$ for each cluster *i* of Ω^B and that the total $\mathbf{X}^B = \sum_{i=1}^{N} \mathbf{X}_i^B$ is known for them. Recall that from (2.6), the estimator $\hat{Y}^B$ can be written as a function of clusters *i* only. From the weights w_i, the *integrated calibration* estimator $\hat{\ddot{Y}}^{CAL,B} = \sum_{i=1}^{n} \ddot{w}_i^{CAL,B} Y_i$ (obtained after using the GWSM) can be determined from the following formulation:

DETERMINE $\ddot{w}_i^{CAL,B}$, FOR $i = 1,\ldots,n$, IN ORDER TO MINIMISE

$$\sum_{i=1}^{n} G_i(\ddot{w}_i^{CAL,B}, w_i) \tag{7.66}$$

UNDER THE CONSTRAINT $\hat{\ddot{\mathbf{X}}}^{CAL,B} = \sum_{i=1}^{n} \ddot{w}_i^{CAL,B} \mathbf{X}_i^B = \mathbf{X}^B$. (7.67)

After having minimised the distance (7.66) under the constraint (7.67), the integrated calibration estimator is obtained:

$$\hat{\ddot{Y}}^{CAL,B} = \sum_{i=1}^{n} \ddot{w}_i^{CAL,B} Y_i = \sum_{i=1}^{n} w_i F_i(\mathbf{X}_i^{T,B} \ddot{\boldsymbol{\lambda}}^B) Y_i \tag{7.68}$$

where $\ddot{w}_i^{CAL,B} = w_i F_i(\mathbf{X}_i^{T,B} \ddot{\boldsymbol{\lambda}}^B)$ is the integrated calibration weight obtained after having used the GWSM. The value of the vector $\ddot{\boldsymbol{\lambda}}^B$ of dimension p^B is the solution of $\mathbf{X}^B = \sum_{i=1}^{n} w_i F_i(\mathbf{X}_i^{T,B} \ddot{\boldsymbol{\lambda}}^B) \mathbf{X}_i^B$.

We can again take as an example the case where the distance function selected is the Euclidean distance $G_i(\ddot{w}_i^{CAL,B}, w_i) = (\ddot{w}_i^{CAL,B} - w_i)^2 / w_i$. With this distance function, we get

$$F_i(\mathbf{x}_i^{T,B} \ddot{\boldsymbol{\lambda}}^B) = 1 + \mathbf{X}_i^{T,B} \ddot{\boldsymbol{\lambda}}^B \tag{7.69}$$

and

$$\ddot{w}_i^{CAL,B} = \ddot{w}_i^{REG,B} = w_i(1 + \mathbf{X}_i^{T,B} \ddot{\boldsymbol{\lambda}}^B) = w_i + w_i \mathbf{X}_i^{T,B} \ddot{\boldsymbol{\lambda}}^B \tag{7.70}$$

with $\ddot{\boldsymbol{\lambda}}^B = \left(\sum_{i=1}^{n} w_i \mathbf{X}_i^B \mathbf{X}_i^{T,B}\right)^{-} (\mathbf{X}^B - \hat{\mathbf{X}}^B)$. The integrated calibration (or integrated regression) estimator obtained with the Euclidean distance thus has the form

$$\hat{\hat{Y}}^{CAL,B} = \hat{\hat{Y}}^{REG,B} = \sum_{i=1}^{n} \ddot{w}_i^{REG,B} Y_i = \hat{Y}^B + (\mathbf{X}^B - \hat{\mathbf{X}}^B)^T \hat{\hat{\boldsymbol{\beta}}}^B \quad (7.71)$$

where $\hat{\hat{\boldsymbol{\beta}}}^B = \left(\sum_{i=1}^{n} w_i \mathbf{X}_i^B \mathbf{X}_i^{T,B}\right)^- \sum_{i=1}^{n} w_i \mathbf{X}_i^B Y_i$. The expression for $\hat{Y}^B$ is given by (2.1) or (4.1), and $\hat{\mathbf{X}}^B$ by (7.21) or (7.22).

With the two estimators $\hat{Y}^{CAL,B}$ and $\hat{\hat{Y}}^{CAL,B}$ having equal weights within each cluster i of Ω^B, the use of calibration before and after the GWSM can now be compared on a common basis. By considering the expression of the integrated calibration weight $\ddot{w}_i^{REG,B}$ given by (7.70), it can be seen that if we set $\ddot{\nabla}_i^{REG,B} = \mathbf{X}_i^B$, we are in the context of generalised calibration of Deville (1998b). Indeed, with the instrumental variable $\ddot{\nabla}_i^{REG,B} = \mathbf{X}_i^B$, the expression (7.70) is brought back exactly to the form that we had in expression (7.16).

We can now go back to this same calibration, but with another instrumental variable. Let us set the following value for the instrumental variable:

$$\nabla_i^{REG,B} = \begin{cases} \dfrac{1}{w_i} \displaystyle\sum_{j=1}^{m^A} d_j^A \frac{L_{j,i}}{L_i^B} \boldsymbol{\Gamma}_j & \text{if } i \in \Omega^B \\ 0 & \text{otherwise.} \end{cases} \quad (7.72)$$

From (7.16), the following development is then obtained:

$$\begin{aligned} w_i^{REG,B} &= w_i + w_i \nabla_i^{T,REG,B} \left(\sum_{i=1}^{n} w_i \nabla_i^{REG,B} \mathbf{X}_i^{T,B}\right)^- (\mathbf{X}^B - \hat{\mathbf{X}}^B) \\ &= w_i + w_i \frac{1}{w_i} \sum_{j=1}^{m^A} d_j^A \frac{L_{j,i}}{L_i^B} \boldsymbol{\Gamma}_j^T \left(\sum_{i=1}^{n} w_i \frac{1}{w_i} \sum_{j=1}^{m^A} d_j^A \frac{L_{j,i}}{L_i^B} \boldsymbol{\Gamma}_j \mathbf{X}_i^{T,B} \right)^- (\mathbf{X}^B - \hat{\mathbf{X}}^B) \\ &= w_i + \sum_{j=1}^{m^A} d_j^A \frac{L_{j,i}}{L_i^B} \boldsymbol{\Gamma}_j^T \left(\sum_{i=1}^{n} \sum_{j=1}^{m^A} d_j^A \frac{L_{j,i}}{L_i^B} \boldsymbol{\Gamma}_j \mathbf{X}_i^{T,B} \right)^- (\mathbf{X}^B - \hat{\mathbf{X}}^B) \\ &= w_i + \sum_{j=1}^{m^A} d_j^A \frac{L_{j,i}}{L_i^B} \boldsymbol{\Gamma}_j^T \left(\sum_{j=1}^{m^A} d_j^A \boldsymbol{\Gamma}_j \sum_{i=1}^{n} \frac{L_{j,i}}{L_i^B} \mathbf{X}_i^{T,B} \right)^- (\mathbf{X}^B - \hat{\mathbf{X}}^B) \\ &= w_i + \sum_{j=1}^{m^A} d_j^A \frac{L_{j,i}}{L_i^B} \boldsymbol{\Gamma}_j^T \left(\sum_{j=1}^{m^A} d_j^A \boldsymbol{\Gamma}_j \boldsymbol{\Gamma}_j^T \right)^- (\mathbf{X}^B - \hat{\mathbf{X}}^B). \end{aligned} \quad (7.73)$$

The last line follows directly from the definition of $\boldsymbol{\Gamma}_j$ given in (7.22).

By comparing (7.73) and (7.53), we notice that the two calibration weights are exactly the same. Thus, with the Euclidean distance, the calibration weight $w_i^{REG,B}$ obtained before the GWSM is the same as the generalised calibration weight obtained after the GWSM with the instrumental variable $\nabla_i^{REG,B}$ given by (7.72).

The two estimators $\hat{Y}^{REG,B}$ and $\hat{\hat{Y}}^{REG,B}$ can therefore be seen as stemming from two generalised calibration estimators carried out after the GWSM with different instrumental variables. It is concluded that the two estimators are not asymptotically equivalent and, as a result, the estimators $\hat{Y}^{CAL,B}$ and $\hat{\hat{Y}}^{CAL,B}$ are not asymptotically equivalent.

In short, performing calibration before or after the use of the GWSM produces different estimators. The question then is to know the extent of this difference. It is this that we are looking to determine in the following section using a simulation study.

7.4.4 Simulation study

We conducted a small simulation study in order to compare the estimators $\hat{Y}^{REG,B}$ and $\hat{\hat{Y}}^{REG,B}$. First, although the estimators are both asymptotically unbiased, they have a certain bias in cases with small or medium-sized samples. On the other hand, these estimators have different precisions that can be interesting to quantify in a manner of knowing which is the most precise.

The study has been inspired by the production of the Whole Farm Data Base of Statistics Canada. This database contains information on livestock, crops and the income and expenditures (tax data) of Canadian farms (Statistics Canada, 2000a). The data used for the simulations come from the agricultural sector for two Canadian provinces: Québec and New Brunswick. The first can be considered a large province, while the second can be seen as a small province. The variable of interest y is the gross farm income, while the auxiliary variable x is the net farm income.

The population U^A is a list of M^A farms coming from the 1996 Farm Register. This register essentially comes from the 1991 Canadian Census of Agriculture, with different updates taking place since then. The units j of U^A thus represent the farms, but note that each farm j can have many farmers. The target population U^B is a list of M^B tax records (or tax reports) from the Canada Revenue Agency (CRA). This second list is the 1996 file of unincorporated businesses from CRA that contains taxl data for people declaring at least agricultural income. The units k are therefore tax reports that are filled out by the different members of a household i (or cluster). The target population U^B has N households. The respective sizes of populations U^A and U^B are given in Table 7.1.

Table 7.1: *Files from Québec and New Brunswick*

	Québec	**New Brunswick**
Size of the Farm Register (U^A)	43 017	4 930
Size of the tax report file (U^B)	52 394	5 155
Number of households in U^B	22 387	2 194
Gross farm income (Y^B)	5 543 853 688	335 989 609

The populations U^A and U^B are related by complex links. Indeed, there are cases where a farm j has many farmers and where each farmer files a tax report k to CRA. We then have a "one-to-many" link since we have a farm j linked to many tax reports k. On the other hand, a farmer who works on more than one farm j can file a single tax report k for the group of farms on which he works. Hence, this is a many-to-one link since there are many farms j linked to a single tax report k. Finally, there are complex links where farmers work on more than one farm and where each farm has a different number of farmers. The populations U^A and U^B, as well as their links, can be represented by Figure 2.1. Note that the links between units j of U^A and ik of U^B were obtained by record linkage. This process will be described in detail in Chapter 9.

A sample s^A of m^A farms is selected from the population U^A according to a certain sampling design. Suppose that π_j^A represents the selection probability of farm *j*. We have $\pi_j^A > 0$ for all farms $j \in U^A$. For each farm *j* selected in s^A, the tax reports *ik* from U^B are identified that have a non-zero link with farm *j*, i.e., $l_{j,ik} = 1$. For each tax report *ik* identified, the list of M_i^B tax reports for the people from household *i* containing this identified tax report is established. Let Ω^B be the set of *n* households identified by the farms $j \in s^A$.

We are interested in estimating the total gross farm income Y^B, which is the income from farming and earned by the members of the households (or clusters) from the target population U^B. To obtain this income, we have the tax reports for all members of the households from Ω^B.

We could question the reason to use a sample of farms from U^A to obtain the tax reports from U^B, instead of simply selecting a sample from U^B, or even directly using the set of data from the population U^B. First, although the data from U^B are available for the entire population, these data require some processing (edit, imputation, etc.) in order to be usable for estimation. As this treatment entails non-negligible costs, it is then necessary to start with a sample instead of a census.[2] A certain advantage can also be drawn in allowing this sample of tax reports to be linked to a sample of farms. Indeed, for the production of statistics on crops and livestock, Statistics Canada conducts a sample of farms. By identifying the tax reports that are links to the farmers owning the farms, the relationships between income and production of crop and livestock can then be studied. It is from the set of data collected on livestock, crops and tax reports that Statistics Canada produces the Whole Farm Data Base (Statistics Canada, 2000a).

---

[2] Note that for the simulations, the entire population was used after a slight processing, assuming that the quality of the data collected was satisfactory.

Although the simulations were performed as inspired by the Whole Farm Data Base, certain processes and data were modified for reasons of confidentiality, and also to avoid needlessly complicating the discussion. However, we believe that these changes do not affect the results from the simulations. The primary objective of the simulations is to compare the two estimators $\hat{Y}^{REG,B}$ and $\hat{\dot{Y}}^{REG,B}$ in an empirical manner, and not to resolve the problems associated with the construction of the Whole Farm Data Base.

For the simulations, the sample s^A from U^A (the Farm Register) is assumed to be selected by simple random sampling without replacement and without any stratification. Two sampling fractions were considered: 30% and 70%. Recall that we are interested in estimating the total gross farm income Y^B, and that a single auxiliary variable x is used, being the net farm income.

Since we have the entire populations of farms and tax reports, it was possible to calculate the value of Y^B, as well as the variances $Var(\hat{Y}^{REG,B})$ and $Var(\hat{\dot{Y}}^{REG,B})$ from the formulas (7.37) and (7.64).

Moreover, because simple random sampling without replacement is assumed, these theoretical formulas could be simplified. Thus, we used

$$Var(\hat{Y}^{REG,B}) \cong M^A \frac{(1-f^A)}{f^A} S_e^2 \tag{7.74}$$

and

$$Var(\hat{\dot{Y}}^{REG,B}) \cong M^A \frac{(1-f^A)}{f^A} S_{\dot{e}}^2 . \tag{7.75}$$

In these expressions:

$f^A = m^A / M^A$ is the *sampling fraction*;

$S_e^2 = \frac{1}{M^A-1} \sum_{j=1}^{M^A} (e_j^A - \bar{e}^A)^2$ with $e_j^A = Z_j - \mathbf{\Gamma}_j^T \boldsymbol{\beta}^B$ and $\bar{e}^A = \sum_{j=1}^{M^A} e_j^A / M^A$;

$S_{\dot{e}}^2 = \frac{1}{M^A-1} \sum_{j=1}^{M^A} (\dot{e}_j^A - \bar{\dot{e}}^A)^2$ with $\dot{e}_j^A = Z_j - \mathbf{\Gamma}_j^T \dot{\boldsymbol{\beta}}^B$ and $\bar{\dot{e}}^A = \sum_{j=1}^{M^A} \dot{e}_j^A / M^A$.

A Monte Carlo study was also performed to calculate in an empirical manner the bias and the variance of $\hat{Y}^{REG,B}$ and $\hat{\hat{Y}}^{REG,B}$. To do this, 1000 samples s^A from U^A were selected for each sampling fraction, 30% and 70%. The empirical bias and the empirical variance of each estimator (represented here by $\hat{Y}$) was calculated using

$$\hat{Bias}(\hat{Y}) = \hat{E}(\hat{Y}) - Y^B = \frac{1}{1000}\sum_{s^A=1}^{1000} \hat{Y}_{s^A} - Y^B \qquad (7.76)$$

and

$$\hat{Var}(\hat{Y}) = \frac{1}{1000}\sum_{s^A=1}^{1000} (\hat{Y}_{s^A} - \hat{E}(\hat{Y}))^2 . \qquad (7.77)$$

The empirical relative bias was calculated from

$$R\hat{Bias}(\hat{Y}) = 100 \times \frac{\hat{Bias}(\hat{Y})}{Y^B} . \qquad (7.78)$$

The Monte Carlo study made it possible to empirically verify (see Table 7.2 below) the accuracy of the asymptotic variance formulas (7.74) and (7.75) obtained for $\hat{Y}^{REG,B}$ and $\hat{\hat{Y}}^{REG,B}$.

Table 7.2: *Simulation results*

Province	f^A	**Statistic**	$\hat{Y}^{REG,B}$	$\hat{\hat{Y}}^{REG,B}$
Québec	0.30	Empirical bias	203 979	2 240 377
		Empirical relative bias (%)	0.004	0.040
		Theoretical variance	2.756×10^{15}	2.700×10^{15}
		Empirical variance	2.623×10^{15}	2.786×10^{15}
	0.70	Empirical bias	–1 108 666	–290 749
		Empirical relative bias (%)	–0.020	–0.005
		Theoretical variance	5.061×10^{14}	4.959×10^{14}
		Empirical variance	5.473×10^{14}	4.814×10^{14}
New Brunswick	0.30	Empirical bias	–722 336	–605 860
		Empirical relative bias (%)	–0.215	–0.180
		Theoretical variance	2.000×10^{14}	2.209×10^{14}
		Empirical variance	2.025×10^{14}	2.161×10^{14}
	0.70	Empirical bias	–345 810	–237 159
		Empirical relative bias (%)	–0.103	–0.071
		Theoretical variance	3.674×10^{13}	4.057×10^{13}
		Empirical variance	3.897×10^{13}	4.076×10^{13}

Looking at the results in Table 7.2, we first notice that the biases of the two estimators $\hat{Y}^{REG,B}$ and $\hat{\hat{Y}}^{REG,B}$ are effectively negligible. Indeed, the largest empirical relative bias in absolute value is 0.215%. From the variance point of view, we notice that there is no estimator that is always better than the other. The difference between the variances of the estimators $\hat{Y}^{REG,B}$ and $\hat{\hat{Y}}^{REG,B}$ is generally not very large at the theoretical or empirical variance level. Furthermore, the theoretical variances turn out to be relatively close to the empirical variances.

Therefore, we finally conclude that, for the Whole Farm Data Base, the estimator $\hat{Y}^{REG,B}$ obtained by calibrating before the GWSM is relatively comparable to the estimator $\hat{\hat{Y}}^{REG,B}$ obtained by calibrating after the GWSM.

CHAPTER 8

NON-RESPONSE

In censuses or sample surveys, it happens inevitably that no value can be obtained for one or several measured variables from certain interviewed units. It is then said that there is *non-response* within the survey. In the case where the values are taken from automated systems, this can include technical problems or breakdowns. On the other hand, when the survey resorts to questionnaires, the values can be missing for different reasons. Examples could include the unwillingness from the surveyed person, gaps in the value asked for, laxity of the interviewer who does not try to obtain responses to all the questions, lost questionnaires, etc. Note that a missing response here, in addition to a loss of information for example, is considered as being a non-response in the same way as a person refusing to respond. On the other hand, if the question allows a "no opinion" option, then this choice is not a non-response.

For certain surveys, it is possible go after missing values by remeasuring or by recontacting the persons surveyed for whom no response has been obtained. However, this recall process often leads to significant costs and delays that cannot always be undertaken by the survey. It is then decided to perform the recall for only a fraction of the non-respondents.

For other surveys, it becomes impossible to redo the measurements or to recontact the non-respondents. Thus, in the case of a sampling of persons, we can come up against a definite refusal, which excludes all possibility of recontact. It can also occasionally happen that the non-responding person is deceased or has moved. Non-response therefore makes the final sample rarely corresponds to the initial sample planned by the survey designers.

An important point to notice is that even if it is physically possible to recontact a non-respondent until a response is obtained, the values obtained will not necessarily be usable by the survey interviewer. Indeed, the quality of the information collected "at any cost" can be so poor that it only contributes toward creating a bias within the estimates. For example, if the surveyed person does not provide a response because he does not know the requested value, forcing him to respond anything is not really useful. In the case of an individual simply refusing to respond, indiscriminate harassment will inevitably lead to a set of erroneous values.

During a recontact, it is essential to remember that it is not just about obtaining a response but to obtain the "correct" response. In many cases, it will be more suitable to treat the non-response with statistical correction rather than trying to fill in the missing values with data containing a significant portion of errors.

There are sizable, though hardly comprehensive, bibliographies in, for example, Droesbeke and Lavallée (1996), Hedges and Olkin (1983), and Bogeström, Larsson and Lyberg (1983).

Since the topic is so broad, we will confine this chapter to *total non-response*, as opposed to *partial non-response*. Total non-response occurs when none of the variables of interest can be measured. For example, the surveyed person simply refused to respond. With partial non-response, only a subset of the variables of interest can be measured. The surveyed person, for example, did not know the answer to one of the questions.

In this chapter, we will study non-response in the context of indirect sampling. Since the GWSM is used in obtaining estimation weights, we will look into the adjustment of these weights to take into account the non-response. The techniques of treating non-response centred on the imputation of missing values will therefore be excluded. On this matter, refer to, among others, Platek and Gray (1983), and Särndal, Swensson and Wretman (1992).

8.1 TYPES OF NON-RESPONSE

With indirect sampling, recall that the selection of a sample s^A from the population U^A is performed in order to produce an estimate for the target population U^B (consisting of clusters) by using the existing correspondence between the two populations. The total non-response here can therefore be present within the sample s^A selected from U^A, or within the units identified to be surveyed within U^B. For example, let us return to the situation illustrated in Figure 1.2 where the target population is children, but where we must first select a sample of parents before we can select the sample of children. Within the sample s^A of parents, there can be people who refuse to give the names of their children for the survey, which creates non-responses within s^A. For the parents who agree to respond, their children can then be identified and the actual survey can proceed. Here also, there will be non-response for the children who refuse, for example, to respond to the survey.

Since the units in population U^B are surveyed by cluster, there are two types of total non-response associated with cluster sampling (direct or indirect): *cluster non-response* and *unit non-response.* Cluster non-response refers to situations where none of the units in the cluster responded to the survey. This is a case often encountered in practice. In telephone surveys, for example, if no one answers the telephone, we then have no response for the entire household (cluster) that we are trying to contact. Moreover, if a person answers the telephone but does not want to participate in the survey, then it is often difficult to obtain a response for the other members of the household.

Unit non-response is a kind of total non-response in which one or more units in the cluster, but not all units, did not respond. Unit non-response also occurs as frequently as cluster non-response, but for other reasons. For example, in a medical survey, a person can respond and describe his own illnesses, but it is not certain that he can answer for all the other members of the household. By contacting a household where many members are absent, there will in all likelihood be unit non-response. Note that some surveys allow for the measurement of variables of interest through an intermediary ("*proxy*"), while others will not ("*non-proxy*"). With a "non-proxy" survey, there can be unit non-response if a unit cannot or does not want to answer the survey questions.

With indirect sampling, there is also another form of non-response that comes from the problem of identifying some of the links. This type of non-response is associated with the situation where it is impossible to determine whether a unit *ik* of U^B is linked to a unit *j* of U^A. This is referred to as the *problem of links identification.* For example, consider the case of longitudinal surveys described in Chapter 6 where the links are one-to-one between populations U^A

(wave 1) and U^B (wave 2). A link is found between populations U^A and U^B through the individuals that belong to both populations. Thus, $l_{j,ik} = 1$ if individual j from population U^A corresponds to individual k of household i from population U^B, and $l_{j,ik} = 0$ otherwise. An individual ik from U^B therefore has a non-zero link with U^A if it was present in the population at wave 1, and a zero link otherwise. During the survey, it can prove to be difficult to know if an individual k from a household i was present or not at wave 1. The individual, for example, can have trouble remembering where he lived at the time of wave 1. Consequently, we cannot know whether or not the individual has a zero link with population U^A, which constitutes a problem of links identification. This kind of non-response problem was already mentioned by Sirken and Nathan (1988) in the context of network sampling. More recently, Ardilly and Le Blanc (2001) faced this problem during the use of the GWSM for the weighting of a survey of homeless people.

Non-response can also be classified into *ignorable* and *non-ignorable* non-responses. Non-response is ignorable when the *response probability* for a certain question, given the selected sample s, does not depend on the value of the variable measured. The fact whether or not a person responds to a question is therefore not related to the response to that question. Let ϕ_k be the probability that person k from the sample s responds to the question measured by the variable of interest y_k. The non-response then is ignorable if $\phi_k = P(\text{unit } k \text{ responds} \mid y_k, s) = P(\text{unit } k \text{ responds} \mid s)$. An example of ignorable non-response is where a questionnaire on employee satisfaction for a company is not returned simply because of negligence. An example of non-ignorable non-response is where only unsatisfied employees return the questionnaire. This last case obviously tends to bias the survey results if no correction is used. To know more about ignorable and non-ignorable non-response, see Rubin (1976), Rubin (1983) and Rubin (1987). We will assume here that the non-response is ignorable.

A final classification of non-response is specific to longitudinal surveys and repeated surveys. For these surveys, we have *attrition* and *wave non-response*. Attrition occurs when a person stops responding for good, beginning with a given wave of interviews. For example, the person cannot be found because he has moved. Note that for deceased persons, they are considered as responding units with the measured variable y_k set to zero. Wave non-response is found when a person does not respond for one or many waves of interview in a temporary manner. For more on this subject, refer to Lepkowski (1989).

In this chapter, treatment of total non-response within the sample s^A will be discussed. Also, we will attempt to treat cluster non-response and unit non-response among those identified to be surveyed within U^B. Finally, we will provide some solutions to the problem of links identification.

8.2 CORRECTING RESPONSE RATES

Response rates take on a particular importance in sample surveys. They can serve, on the one hand, to measure the progress or the performance of the survey collection and, on the other hand, to help correct the estimates taking into account the non-response. Two categories of response rates can then be distinguished: *operational response rates* and *corrective response rates*. These two categories are qualitative rates since they contribute to qualitatively assess the collection results. Operational response rates are so called since they serve to evaluate the quality of survey operations. For example, an operational response rate could be the relationship between the number of interviews completed and the number of persons contacted. The corrective response rates serve more to correct the estimates taking into account the non-response (Droesbeke and Lavallée, 1996).

The corrective response rates can correct the total non-response by drawing the subsample of respondents toward the initial sample. They have a more restrictive meaning than operational response rates. In fact, they must reflect the importance of the number of respondents in the survey in comparison to the initial sample.

In the general context of a population U of size N, a sample s of size n is selected where each unit k is selected with probability $\pi_k > 0$. We attempt to measure a variable of interest y_k where unfortunately non-responses are present. Let n_r be the number of units responding to the survey. The corrective response rate $\dot{R}$ is defined as the ratio between the number of responding units n_r from the sample and the sample size n, i.e.,

$$\dot{R} = \frac{n_r}{n}. \tag{8.1}$$

Using the selection probabilities π_k of the units k from the sample, a weighted version of the corrective response rate can be defined. This latest version is given by

$$R = \frac{\sum_{k=1}^{n_r} 1/\pi_k}{\sum_{k=1}^{n} 1/\pi_k} = \frac{\hat{N}_r}{\hat{N}}. \tag{8.2}$$

The weighted corrective response rate can be seen as the ratio between the estimated number of responding units within the population and the estimated number of units in the population. Although the debate remains open about the use of $\dot{R}$ or R, we will prefer here the weighted version given by (8.2).

A first reason to use the weighted response rate R, instead of $\dot{R}$, is related to the estimate of the size of the population that remains unchanged, whether there is non-response or not. Indeed, if there is no non-response, $\hat{N} = \sum_{k=1}^{n} d_k$ where $d_k = 1/\pi_k$. If there is non-response the sampling weight d_k can be corrected by using the corrective response rate and thus obtaining a new weight d_k^{NR} corrected for the non-response. Starting with the response rate (8.2), $d_k^{NR} = d_k / R$ and then we note that $\sum_{k=1}^{n_r} d_k^{NR} = \hat{N}$. The sum of the sampling weights corrected with the weighted response rate therefore gives the same result as if there was no non-response. However, with the response rate (8.1), $\dot{d}_k^{NR} = d_k / \dot{R}$ and, in this case, $\sum_{k=1}^{n_r} \dot{d}_k^{NR} = n\hat{N}_r / n_r \neq \hat{N}$. Therefore, the desired result is not obtained.

Another reason to use the response rate (8.2) is related to model-based considerations. As mentioned by Pfeffermann (1993), the use of weights in survey data modelling results in parameter estimates that are consistent with respect to the sampling design. In addition, the use of weights reduces the impact of a poor model formulation on the estimates. It will be seen in section 8.3 that the corrective response rate can be seen as an estimate of the response probability ϕ_k of a unit k from the sample. The corrective response rate is therefore the result of modelling the response probability ϕ_k, from which the suggestion of using selection probabilities in its calculation comes. It is finally noted that the two corrective response rates, weighted and unweighted, are equal when the selection probabilities π_k are equal.

The corrective response rates can be calculated for exclusive and exhaustive groups within the population. For example, these Q groups can be the strata, but they can also be the result of any partition of the population. For a group q, the corrective response rate R_q is thus defined as

$$R_q = \frac{\sum_{k=1}^{n_{r,q}} 1/\pi_k}{\sum_{k=1}^{n_q} 1/\pi_k} = \frac{\hat{N}_{r,q}}{\hat{N}_q} \tag{8.3}$$

where n_q is the number of units from the sample belonging to group q and $n_{r,q}$ is the number of responding units belonging to group q. It will be seen that the groups q can be formed in a way that they correspond to sets where the response probabilities of the units included within them are relatively homogeneous.

It is important to note that it is essential, to calculate the response rate (8.3), that information concerning the non-respondents themselves be available. For example, if the groups are formed from the socioprofessional category of persons in the survey, it is then necessary to know the categories of the non-respondents, like those of the respondents. This can represent a problem is the socioprofessional category is measured in the course of the survey.

The people that are not contacted because they are out-of-scope for the survey are part of the respondents (and not the non-respondents) but their variables of interest are all set to zero. This reflects the fact that other people outside of the sample can also be out-of scope for the survey, but these people are often only known during the interview.

8.3 RESPONSE PROBABILITIES

The notion of response probability was briefly touched by presenting, in section 8.1, the aspect of ignorable and non-ignorable non-response. This concept will be developed here in further detail, as presented by Särndal, Swensson and Wretman (1992) and Lock Oh and Scheuren (1983).

The concept of response probability is very useful in adjusting estimates for total non-response. In a general context, let δ_k be an indicator variable that takes a value of 1 if unit k answers the survey questions, and 0 if not. It is generally assumed that this variable has a

Bernoulli distribution with probability ϕ_k. In other words, it is assumed that each individual k in the survey population has a certain probability ϕ_k of responding to the survey, i.e., $P(\text{unit } k \text{ reponds} \mid s) = P(\delta_k = 1 \mid s) = \phi_k$. In addition, for two units k and k', the indicator variables δ_k and $\delta_{k'}$ are deemed to be independent. This implies that the joint probability of response $\phi_{kk'}$ for these two units is given by

$$\begin{aligned}\phi_{kk'} &= P(\delta_k = 1, \delta_{k'} = 1 \mid s)\\ &= P(\delta_k = 1 \mid s) P(\delta_{k'} = 1 \mid s) = \phi_k \phi_{k'}.\end{aligned}$$

Lastly, we have

$$\begin{aligned}E(\delta_k \mid s) &= 1 \times P(\delta_k = 1 \mid s) + 0 \times P(\delta_k = 0 \mid s)\\ &= P(\delta_k = 1 \mid s) = \phi_k\end{aligned} \tag{8.4}$$

and

$$\begin{aligned}Var(\delta_k \mid s) &= E(\delta_k^2 \mid s) - E^2(\delta_k \mid s)\\ &= E(\delta_k \mid s) - E^2(\delta_k \mid s)\\ &= \phi_k - \phi_k^2 = \phi_k(1 - \phi_k).\end{aligned} \tag{8.5}$$

The independence between the indicator variables δ of two units k and k' follows from the assumption that the choice made by unit k to respond or not will have no bearing on the choice made by unit k'. In other words, there is no ratchet effect. This turns out generally to be true in practice, except in the case of cluster sampling where the *cluster effect* (or *intracorrelation*) can nullify this independence. For example, if the units are individuals selected from a cluster sampling, we can imagine that the fact that one of the individuals responded to the survey could prompt other individuals from the cluster to respond, and the opposite can also be true. Unless otherwise informed, the assumption of independence will be kept here in order to simplify the discussion.

As we have above, the response probability can depend on the sample. Thus, a unit k can have more of a tendency to respond for a certain sample s compared to another sample s'. For example, in a sequential sample of a group of persons placed in order, a person chosen last can be less inclined to respond than a person chosen first simply because he has been waiting for a longer time.

The estimation of the response probabilities ϕ_k can be done with different approaches. From the information provided by the survey and by external sources, we normally seek to develop a model that is meant to identify the factors influencing the response probabilities. This model can take forms ranging from very simple to relatively complex. For example, it can be determined that the probabilities ϕ_k for an employee satisfaction survey are uniquely influenced by the sex of the persons surveyed. An estimate $\hat{\phi}_k$ of the response probabilities ϕ_k is then simply given by the corrective response rate observed in the survey for each of the two sexes, i.e., $\hat{\phi}_k = n_{r,\text{man}} / n_{\text{man}}$ if unit k is a man, and $\hat{\phi}_k = n_{r,\text{woman}} / n_{\text{woman}}$ if unit k is a woman. Recall that, to estimate the response probabilities, it is essential to use information concerning the non-respondents themselves. In the previous example, we see that the sex was used as auxiliary information.

In a general way, we can try to estimate the response probabilities by calculating the corrective response rates within *response homogeneity groups* (RHG) as suggested by Särndal, Swensson and Wretman (1992). The RHG form in fact a partition of the sample into Q groups, where the response probabilities of the units from the sample are approximately the same within each group q. They can be represented by the model

$$\phi_{qk} = E(\delta_{qk} \mid s) = \beta_q , \tag{8.6}$$

where β_q is a fixed effect (or factor) to be estimated. The parameter β_q is in fact the expected probability of response in group q. The RHG can be formed by a single factor or by a combination of two or more factors. For example, we can think of age-sex groups.

To estimate ϕ_{qk}, we can use the *weighted maximum likelihood method* (Collett, 1991) with weights set at $d_k = 1/\pi_k$. The following quantity is then maximised:

$$\ln \Lambda(\beta_1,...,\beta_Q) = \sum_{q=1}^{Q} \sum_{k=1}^{n_q} d_{qk} [\delta_{qk} \ln \phi_{qk} + (1-\delta_{qk}) \ln(1-\phi_{qk})]$$

where the ϕ_{qk} depend on the parameters β_q. The ensuing estimator from the model (8.6) is then given by the corrective response rate (8.3). We thus have

$$\hat{\phi}_{qk} = R_q . \tag{8.7}$$

Another approach used to estimate the response probabilities consists of using a *logistic regression model* (or *"logit" model*). In a general way, the logistic regression model is given by

$$\log\left(\frac{\phi_k}{1-\phi_k}\right)=\boldsymbol{\beta}^T\mathbf{x}_k, \tag{8.8}$$

where $\boldsymbol{\beta}$ is a column vector of dimension p of parameters to be estimated, and $\mathbf{x}_k$ is a column vector of auxiliary variables. The vector of parameters $\boldsymbol{\beta}$ is estimated, again using the weighted maximum likelihood method. With the logistic regression model, we obtain an estimate $\hat{\boldsymbol{\beta}}$ of $\boldsymbol{\beta}$, from which the following estimates are derived:

$$\hat{\phi}_k^{LOGIT}=\frac{\exp(\hat{\boldsymbol{\beta}}^T\mathbf{x}_k)}{1+\exp(\hat{\boldsymbol{\beta}}^T\mathbf{x}_k)}. \tag{8.9}$$

It is interesting to note that if the auxiliary variables $\mathbf{x}_k$ are all qualitative and if the model chosen is saturated (that is, that there are as many parameters to estimate as there are combinations of values), then the logistic regression model approach corresponds exactly to that of the RHG.

If the number of factors explaining the response probabilities is large, it is in practice simpler to use the logistic regression approach because the RHG approach requires making combinations for all the factors influencing the response probabilities, which can result in the creation of groups q without respondents $(n_{r,q}=0)$. The logistic regression approach only requires that the marginal response rates correspond to the factors present in the model chosen. Michaud and Hunter (1992) used it in order to determine the decisive non-response factors in SLID. Starting from the generalised calibration theory presented in section 7.1, Deville (1998b) (and Deville, 2000b) demonstrated that the method used to estimate the response probabilities is only of little importance. Indeed, from the two estimates corrected with the different response probabilities, we see that the difference between these two estimates is asymptotically zero.

8.4 TREATMENT OF NON-RESPONSE WITHIN s^A

Non-response in sample s^A is a classic case of non-response. Whether in the context of conventional (or direct) sampling or indirect sampling, the treatment of this type of non-response is covered in most books on sampling theory. In theorem 4.1, we saw that the estimator $\hat{Y}^B$ produced by the GWSM can be written in the form of a Horvitz-Thompson estimator that is a function of units *j* of s^A. Hence, non-response in s^A is treated as we would treat non-response in the situation where we selected sample s^A to produce an estimate of a quantity related to population U^A. We still present here the treatment of this type of non-response because it will allow us to establish the basis of the discussion for the other types of non-response presented in this chapter.

A sample s^A is selected containing m^A units from the population U^A consisting of M^A units according to a certain sampling design. Suppose that π_j^A represents the selection probability of unit *j*. We assume that $\pi_j^A > 0$ for all $j \in U^A$. It is assumed that a subset s_r^A of m_r^A units of sample s^A answered the survey questions. It is also assumed that there is only total non-response here and no partial non-response. This situation is illustrated in Figure 8.1. The arrows indicate that units $j = 1$ and $j = 2$ from U^A were selected to be part of s^A. Unit $j = 2$ answered the survey, but unit $j = 1$ did not.

The target population U^B contains M^B units. This population is divided into *N* clusters, where cluster *i* contains M_i^B units. For each unit *j* of s_r^A, units *ik* from U^B can be identified that have a non-zero link $l_{j,ik}$ with *j*, i.e., $l_{j,ik} = 1$. For each identified unit *ik*, we assume that we can make a list of the M_i^B units of cluster *i* containing that unit. Each cluster *i* represents, then, by itself, a population U_i^B where $U^B = \bigcup_{i=1}^{N} U_i^B$. Let Ω_r^B be the set of n_r clusters identified by the units $j \in s_r^A$.

We survey all units *k* of clusters $i \in \Omega_r^B$, where we measure the variable of interest *y*. For target population U^B, we want to estimate the total $Y^B = \sum_{i=1}^{N} \sum_{k=1}^{M_i^B} y_{ik}$. It is assumed that we have the total number of links L_i^B for each cluster $i \in \Omega_r^B$.

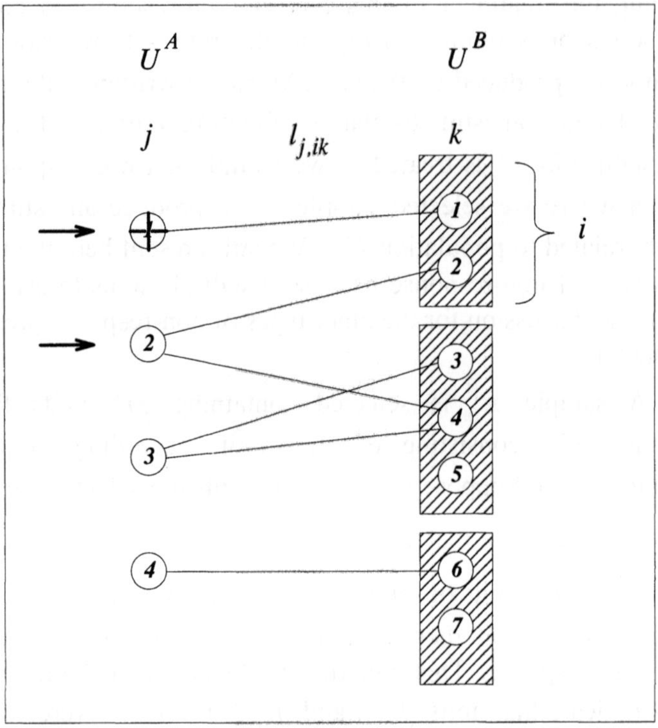

Figure 8.1: *Example of non-response within* s^A

In applying the GWSM, we want to assign an estimation weight w_{ik} to each unit k of surveyed cluster i. To estimate the total Y^B for target population U^B, then, we can use the estimator (2.1), which was constructed on the assumption that there is no non-response in sample s^A. On the basis of Theorem 4.1, we can rewrite the estimator (2.1) as (4.1), which is a function of units j of s^A. Since we have only subsample s_r^A of the responding units, we have to use an estimator that has been corrected for non-response. To that end, we can use the following estimator:

$$\hat{Y}^{NRA,B} = \sum_{j=1}^{M^A} \frac{t_j \delta_j^A}{\pi_j^A \phi_j^A} Z_j = \sum_{j=1}^{m_r^A} \frac{Z_j}{\pi_j^A \phi_j^A} \tag{8.10}$$

where ϕ_j^A is the response probability of unit j. The superscript "NRA" refers to the non-response within s^A. The indicator variable $\delta_j^A = 1$ if unit j of s^A responds, and 0 if not. The probability ϕ_j^A can depend on the sample s^A. Let $\boldsymbol{E}_s(\cdot)$ denote the expected value carried out in relation to all possible samples of s^A. To show that this estimator is unbiased, we proceed as follows:

$$
\begin{aligned}
E(\hat{Y}^{NRA,B}) &= E_s[E(\hat{Y}^{NRA,B} \mid s^A)] = E_s\left[\sum_{j=1}^{M^A} \frac{t_j E(\delta_j^A \mid s^A)}{\pi_j^A \phi_j^A} Z_j\right] \\
&= E_s\left[\sum_{j=1}^{M^A} \frac{t_j \phi_j^A}{\pi_j^A \phi_j^A} Z_j\right] = E_s\left[\sum_{j=1}^{M^A} \frac{t_j}{\pi_j^A} Z_j\right] \qquad (8.11) \\
&= Z = Y^B
\end{aligned}
$$

as, from (8.4), $E(\delta_j^A \mid s^A) = \phi_j^A$. The last line follows directly from Corollary 4.1.

Note that the quantity $1/\pi_j^A \phi_j^A$ corresponds to the sampling weight adjusted to account for non-response. From this adjusted sampling weight, the estimation weight w_{ik}^{NRA} can then be obtained following steps 1 to 4 of the GWSM described in section 2.1.

Steps of the GWSM adjusted for non-response within s^A

Step 1: For each unit k of clusters i from Ω_r^B, the initial weight $w_{ik}'^{NRA}$ is calculated, to know:

$$
w_{ik}'^{NRA} = \sum_{j=1}^{M^A} l_{j,ik} \frac{t_j \delta_j^A}{\pi_j^A \phi_j^A}. \qquad (8.12)
$$

Step 2: For each unit k of clusters i from Ω_r^B, the total number of links $L_{ik}^B = \sum_{j=1}^{M^A} l_{j,ik}$ is obtained.

Step 3: The final weight w_i^{NRA} is calculated:

$$
w_i^{NRA} = \frac{\sum_{k=1}^{M_i^B} w_{ik}'^{NRA}}{\sum_{k=1}^{M_i^B} L_{ik}^B}. \qquad (8.13)
$$

Step 4: Finally, we set $w_{ik}^{NRA} = w_i^{NRA}$ for all $k \in U_i^B$.

After applying the non-response-adjusted GWSM, we can assign an estimation weight w_{ik}^{NRA} to each unit k of the n_r surveyed clusters. To estimate the total Y^B for the target population U^B, the following estimator is then used:

$$\hat{Y}^{NRA,B} = \sum_{i=1}^{n_r} \sum_{k=1}^{M_i^B} w_{ik}^{NRA} y_{ik} . \tag{8.14}$$

In practice, estimator $\hat{Y}^{NRA,B}$ is useful only if the value of the response probabilities ϕ_j^A is known for all units j of s_r^A. We want to estimate these probabilities so that we can use one of the following forms:

$$\hat{\hat{Y}}^{NRA,B} = \sum_{j=1}^{m_r^A} \frac{Z_j}{\pi_j^A \hat{\phi}_j^A} \tag{8.15a}$$

or

$$\hat{\hat{Y}}^{NRA,B} = \sum_{i=1}^{n_r} \sum_{k=1}^{M_i^B} \hat{w}_{ik}^{NRA} y_{ik} , \tag{8.15b}$$

where the weight $\hat{w}_{ik}^{NRA}$ is obtained by replacing ϕ_j^A with $\hat{\phi}_j^A$ in (8.12). To obtain $\hat{\phi}_j^A$, either the estimator (8.7) or the estimator (8.9) can be used.

Here, the model (8.6) takes the form: $\phi_{qj}^A = \beta_q^A$. In view of the estimator (8.7) based on this model, we use the weighted response rate $R_q^A = (\sum_{j=1}^{m_{r,q}^A} 1/\pi_{qj}^A)/(\sum_{j=1}^{m_q^A} 1/\pi_{qj}^A)$.

Thus we have

$$\begin{aligned} \hat{\hat{Y}}^{NRA,B} &= \sum_{j=1}^{m_r^A} \frac{Z_j}{\pi_j^A \hat{\phi}_j^A} = \sum_{q=1}^{Q} \sum_{j=1}^{m_{r,q}^A} \frac{Z_{qj}}{\pi_{qj}^A R_q^A} \\ &= \sum_{q=1}^{Q} \frac{\sum_{j=1}^{m_{r,q}^A} Z_{qj} / \pi_{qj}^A}{\sum_{j=1}^{m_{r,q}^A} 1/\pi_{qj}^A} \left(\sum_{j=1}^{m_q^A} 1/\pi_{qj}^A \right). \end{aligned} \tag{8.16}$$

It is assumed here that the number of responding units $m_{r,q}^A$ is greater than 0 for all RHG q.

If we look at estimator (8.16), we see that it is nothing more than a ratio estimator in two-phase sampling. Särndal, Swensson and Wretman (1992) present a proof that estimator (8.16) is asymptotically unbiased, under the conditions of model (8.6). The asymptotic variance of $\hat{\hat{Y}}^{NRA,B}$ is calculated using a conditional approach. From the identity

$$Var(\hat{\hat{Y}}^{NRA,B}) = Var_s[E(\hat{\hat{Y}}^{NRA,B} \mid s^A)] + E_s[Var(\hat{\hat{Y}}^{NRA,B} \mid s^A)]$$

we get

$$Var(\hat{\hat{Y}}^{NRA,B}) \cong \sum_{j=1}^{M^A}\sum_{j'=1}^{M^A} \frac{(\pi_{jj'}^A - \pi_j^A \pi_{j'}^A)}{\pi_j^A \pi_{j'}^A} Z_j Z_{j'}$$
$$+ E_s\left[\sum_{q=1}^{Q} \frac{(1-\beta_q^A)}{\beta_q^A} \sum_{j=1}^{m_q^A} \frac{1}{(\pi_{qj}^A)^2}\left(Z_{qj} - \frac{\sum_{j=1}^{m_q^A} Z_{qj}/\pi_{qj}^A}{\sum_{j=1}^{m_q^A} 1/\pi_{qj}^A}\right)^2 \Bigg| s^A\right]$$

(8.17)

where β_q^A is the parameter from model (8.6). The variance (8.17) can be estimated using

$$\hat{V}ar(\hat{\hat{Y}}^{NRA,B}) = \sum_{j=1}^{m_r^A}\sum_{j'=1}^{m_r^A} \frac{(\pi_{jj'}^A - \pi_j^A \pi_{j'}^A)}{\pi_{jj'}^A \pi_j^A \hat{\phi}_j^A \pi_{j'}^A \hat{\phi}_{j'}^A} Z_j Z_{j'}$$
$$+ \sum_{q=1}^{Q} \frac{(1-R_q^A)}{(R_q^A)^2} \sum_{j=1}^{m_{r,q}^A} \frac{1}{(\pi_{qj}^A)^2}\left(Z_{qj} - \frac{\sum_{j=1}^{m_{r,q}^A} Z_{qj}/\pi_{qj}^A}{\sum_{j=1}^{m_{r,q}^A} 1/\pi_{qj}^A}\right)^2.$$

(8.18)

If we have auxiliary variables $\mathbf{x}_j^A$ available for all units j of s^A, we can imagine using the model (8.8) with parameter $\boldsymbol{\beta}^A$ of dimension p^A. With the estimator (8.9) based on this model, we then have:

$$\hat{\hat{Y}}^{NRA,LOGIT,B} = \sum_{j=1}^{m_r^A} \frac{Z_j}{\pi_j^A \hat{\phi}_j^{LOGIT,A}} = \sum_{j=1}^{m_r^A} \frac{Z_j}{\pi_j^A} \frac{(1+\exp\ (\hat{\boldsymbol{\beta}}^{T,A}\ \mathbf{x}_j^A))}{\exp\ (\hat{\boldsymbol{\beta}}^{T,A}\ \mathbf{x}_j^A)},$$

(8.19)

where the estimator $\hat{\boldsymbol{\beta}}^A$ is obtained using the weighted maximum likelihood method, with the weights set at $d_j = 1/\pi_j^A$. Note that the estimator $\hat{\hat{Y}}^{NRA,LOGIT,B}$ is highly nonlinear. It is not simple to calculate its bias. However, if the estimate of $\hat{\phi}_j^{LOGIT,A}$ turns out to be relatively close to the true response probability ϕ_j^A, this bias should be small. It is possible to obtain an approximation of its variance by using Taylor linearisation, from which an estimate of the variance of $\hat{\hat{Y}}^{NRA,LOGIT,B}$ is subsequently obtained. Such an approach, however, is rarely used in practice at Statistics Canada, where the Jackknife and Bootstrap methods are preferred instead. If the sampling design used to select the sample s^A is a stratified multi-stage design, the Jackknife estimator of the variance of $\hat{\hat{Y}}^{NRA,LOGIT,B}$ has the form (6.9). To learn more about the Jackknife method, refer to Wolter (1985), and Särndal, Swensson and Wretman (1992).

8.5 TREATMENT OF CLUSTER NON-RESPONSE

As mentioned in section 8.1, cluster non-response occurs when no units of a cluster from U^B identified to be surveyed responds to the survey. This is a case frequently encountered in practice. In this section, the treatment of this type of non-response is presented. As in section 8.4, it is proposed here to treat this type of non-response by using the concept of response probability.

A sample s^A is again selected containing m^A units from the population U^A consisting of M^A units according to a certain sampling design. Let $\pi_j^A > 0$ represent the selection probability of unit j. Contrary to section 8.4, it is assumed that the set of m^A units from the sample responded to the survey questions.

The target population U^B contains M^B units. This population is divided into N clusters, where cluster i contains M_i^B units. For each unit j selected in s^A, we identify the units ik of U^B that have a non-zero relationship $l_{j,ik}$ with j, i.e., $l_{j,ik} = 1$. For each unit ik identified, it is assumed that a list of the M_i^B units of cluster i containing that unit can be made. Each cluster i represents, then, by itself, a population U_i^B, where

$$U^B = \bigcup_{i=1}^{N} U_i^B .$$

Let Ω^B be the set of n clusters identified by the units $j \in s^A$.

In carrying out the survey process, we attempt to survey all units k in clusters i of Ω^B. Unfortunately, for some entire clusters, we are unable to obtain any data. This is cluster non-response. We assume here that all the units of each cluster i from Ω^B respond or do not respond. In other words, there are no clusters in which only a non-zero subset of units responded. Let Ω_r^B be the set of n_r responding clusters. It is worth noting that Ω_r^B differs in general to the set of responding clusters in the context of non-response within the sample s^A. This situation is illustrated in Figure 8.2. The arrows indicate that units $j=1$ and $j=2$ from U^A were selected to be part of s^A. Then, clusters $i=1$ and $i=2$ are identified to be surveyed. Only cluster $i=2$ here answers the survey.

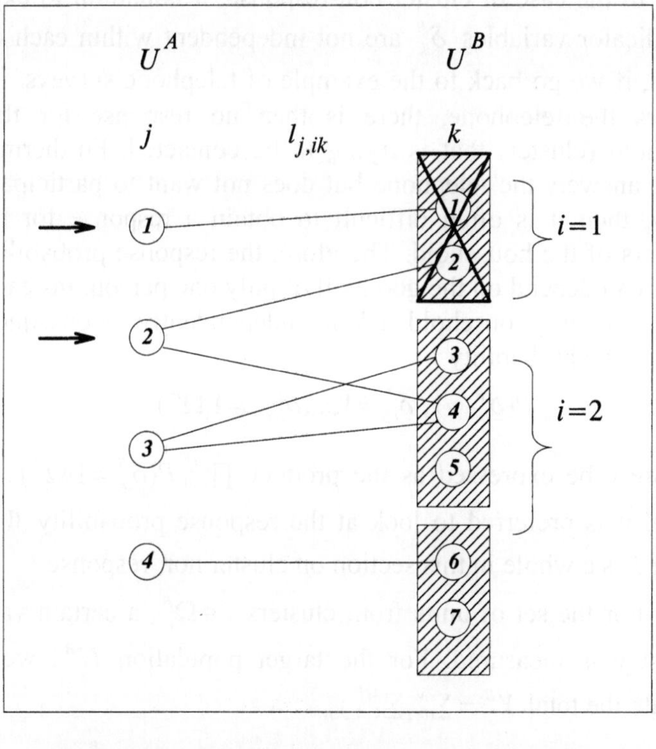

Figure 8.2: *Example of cluster non-response*

Let δ_i^B be an indicator variable that takes a value of 1 if cluster i answers the survey questions, and 0 if not. As in section 8.3, it is assumed that this variable has a *Bernoulli distribution* with probability Φ_i^B. In other words, it is assumed that each cluster i in U^B has a probability Φ_i^B of responding to the survey, i.e.,

$$P(\text{cluster } i \text{ responds} \mid \Omega^{\mathrm{B}}) = P(\delta_i^B = 1 \mid \Omega^{\mathrm{B}}) = \Phi_i^B .$$

In addition, for two clusters i and i', the indicator variables δ_i^B and $\delta_{i'}^B$ are deemed to be independent.

It is worth noting that it is also possible to define the response probability Φ_i^B from the indicator variables δ_{ik}^B associated with the units k of the surveyed clusters i. Let $\delta_{ik}^B = 1$ if unit k of cluster i answers the survey questions, and 0 otherwise. The response probability of cluster i can be defined as being the probability that all the units of the cluster respond, i.e., $\Phi_i^B = P(\delta_{i1}^B = 1,\ \delta_{i2}^B = 1,\ldots,\delta_{iM_i^B}^B = 1|\Omega^B)$.

In the case of cluster non-response, it is natural to expect that the indicator variables δ_{ik}^B are not independent within each cluster i. Indeed, if we go back to the example of telephone surveys, if no one answers the telephone, there is then no response for the entire household (cluster) that is trying to be contacted. Furthermore, if a person answers the telephone but does not want to participate in the survey, then it is often difficult to obtain a response for the other members of the household. Therefore, the response probability of the cluster can depend on the goodwill of only one person, instead of each person of the household taken independently. Consequently, in practice, the probability

$$P(\delta_{i1}^B = 1,\ \delta_{i2}^B = 1,\ldots,\delta_{iM_i^B}^B = 1|\Omega^B)$$

can rarely be expressed as the product $\prod_{k=1}^{M_i^B} P(\delta_{ik}^B = 1|\Omega^{\mathrm{B}})$. For this reason, it is preferred to look at the response probability Φ_i^B of the cluster i as a whole in this section on cluster non-response.

For the set of units from clusters $i \in \Omega_r^B$, a certain variable of interest y is measured. For the target population U^B, we look to estimate the total $Y^B = \sum_{i=1}^{N} \sum_{k=1}^{M_i^B} y_{ik}$.

In applying the GWSM, we want to assign an estimation weight w_{ik}^{NRC} to each unit k of responding cluster i. The superscript "NRC" refers to the non-response of the clusters. To estimate the total Y^B for target population U^B, we can then use the estimator

$$\hat{Y}^{NRC,B} = \sum_{i=1}^{n_r}\sum_{k=1}^{M_i^B} w_{ik}^{NRC} y_{ik} = \sum_{i=1}^{n} \delta_i^B \sum_{k=1}^{M_i^B} w_{ik}^{NRC} y_{ik} . \quad (8.20)$$

To obtain the weight w_{ik}^{NRC} from the GWSM, we are going to use the response probability Φ_i^B for each cluster $i \in \Omega_r^B$.

Steps of the GWSM adjusted for cluster non-response

Step 1: For each unit k of the clusters i from Ω_r^B, the initial weight w'_{ik} is calculated, to know:

$$w'_{ik} = \sum_{j=1}^{M^A} l_{j,ik} \frac{t_j}{\pi_j^A}, \quad (8.21)$$

where $t_j = 1$ if $j \in s^A$, and 0 otherwise.

Step 2: For each unit k of the clusters i from Ω_r^B, the number of total links is obtained:

$$L_{ik}^B = \sum_{j=1}^{M^A} l_{j,ik} . \quad (8.22)$$

Step 3: The final weight w_i^{NRG}, adjusted for non-response, is calculated:

$$w_i^{NRC} = \frac{1}{\Phi_i^B} \frac{\sum_{k=1}^{M_i^B} w'_{ik}}{\sum_{k=1}^{M_i^B} L_{ik}^B} . \quad (8.23)$$

Step 4: Finally, we set $w_{ik}^{NRC} = w_i^{NRC}$ for all $k \in U_i^B$, for all clusters i from Ω_r^B.

Note that for each unit k of the clusters i from Ω_r^B, we have

$$w_{ik}^{NRC} = w_i^{NRC} = \frac{1}{\Phi_i^B} w_{ik} , \quad (8.24)$$

where w_{ik} is the weight from the GWSM without cluster non-response. Furthermore, from Result 2.1, we get

$$w_{ik}^{NRC} = \frac{1}{\Phi_i^B} w_{ik} = \sum_{j=1}^{M^A} \frac{t_j}{\pi_j^A} \frac{L_{j,i}}{\Phi_i^B L_i^B}. \tag{8.25}$$

Let $E_s(\cdot)$ denote the expected value carried out for all possible samples of s^A where each sample s^A leads, as we recall, to a set of clusters Ω^B. To show that the estimator $\hat{Y}^{NRC,B}$ is unbiased, we proceed from a conditional approach.

From (8.20), we have

$$\begin{aligned} E(\hat{Y}^{NRC,B}) &= E_s[E(\hat{Y}^{NRC,B} \mid \Omega^B)] = E_s\left[\sum_{i=1}^{n} E(\delta_i^B \mid \Omega^B) \sum_{k=1}^{M_i^B} w_{ik}^{NRC} y_{ik}\right] \\ &= E_s\left[\sum_{i=1}^{n} \Phi_i^B \sum_{k=1}^{M_i^B} w_{ik}^{NRC} y_{ik}\right] = E_s\left[\sum_{i=1}^{n} \Phi_i^B \sum_{k=1}^{M_i^B} \frac{w_{ik}}{\Phi_i^B} y_{ik}\right] \\ &= E_s\left[\sum_{i=1}^{n}\sum_{k=1}^{M_i^B} w_{ik} y_{ik}\right] = E_s\left[\hat{Y}^B\right] = Y^B. \end{aligned} \tag{8.26}$$

The last line follows directly from Corollary 4.1.

Theorem 8.1: Duality of the form of $\hat{Y}^{NRC,B}$

Let $Y_i = \sum_{k=1}^{M_i^B} y_{ik}$ and $L_i^B = \sum_{k=1}^{M_i^B} L_{ik}^B$ for all $k \in U_i^B$. The estimator $\hat{Y}^{NRC,B}$ given by (8.20) can then also be written under the form

$$\hat{Y}^{NRC,B} = \sum_{j=1}^{M^A} \frac{t_j}{\pi_j^A} Z_j^{NRC} \tag{8.27}$$

where

$$Z_j^{NRC} = \sum_{i=1}^{n} \frac{L_{j,i}}{L_i^B} \frac{\delta_i^B}{\Phi_i^B} Y_i . \tag{8.28}$$

Proof

From $\hat{Y}^{NRC,B} = \sum_{i=1}^{n_r} w_i^{NRC} \sum_{k=1}^{M_i^B} y_{ik} = \sum_{i=1}^{n_r} w_i^{NRC} Y_i$, we use identity (8.25) to get

$$\begin{aligned} \hat{Y}^{NRC,B} &= \sum_{i=1}^{n_r} Y_i \left(\sum_{j=1}^{M^A} \frac{t_j}{\pi_j^A} \frac{L_{j,i}}{\Phi_i^B L_i^B} \right) \\ &= \sum_{i=1}^{n_r}\sum_{j=1}^{M^A} \frac{t_j}{\pi_j^A} \frac{1}{\Phi_i^B} \frac{L_{j,i}}{L_i^B} Y_i = \sum_{i=1}^{n}\sum_{j=1}^{M^A} \frac{t_j}{\pi_j^A} \frac{\delta_i^B}{\Phi_i^B} \frac{L_{j,i}}{L_i^B} Y_i . \end{aligned} \tag{8.29}$$

By continuing the development, we get

$$\hat{Y}^{NRC,B} = \sum_{i=1}^{n}\sum_{j=1}^{M^A}\frac{t_j}{\pi_j^A}\frac{\delta_i^B}{\Phi_i^B}\frac{L_{j,i}}{L_i^B}Y_i$$
$$= \sum_{j=1}^{M^A}\frac{t_j}{\pi_j^A}\sum_{i=1}^{n}\frac{\delta_i^B}{\Phi_i^B}\frac{L_{j,i}}{L_i^B}Y_i = \sum_{j=1}^{M^A}\frac{t_j}{\pi_j^A}Z_j^{NRC}. \tag{8.30}$$

■

The estimator $\hat{Y}^{NRC,B}$ can thus be written as a function of units *ik* from U^B, or as a function of units *j* from U^A. Note that contrary to the quantity Z_j defined by (4.2), the quantity Z_j^{NRC} defined by (8.28) depends on the set Ω^B of clusters that can be surveyed, and therefore of the sample s^A through the variable δ_i^B.

In practice, the estimator $\hat{Y}^{NRC,B}$ is only useful if the value of the response probabilities Φ_i^B is known for all clusters *i* from Ω_r^B. We then want to estimate these probabilities so that we can use the following estimator:

$$\hat{\hat{Y}}^{NRC,B} = \sum_{i=1}^{n_r}\sum_{k=1}^{M_i^B}\hat{w}_{ik}^{NRC}y_{ik}, \tag{8.31}$$

where $\hat{w}_{ik}^{NRC} = w_{ik} / \hat{\Phi}_i^B$. To obtain $\hat{\Phi}_i^B$, we can follow the example of estimator (8.7) or of estimator (8.9). In this case, model (8.6) takes the form: $\Phi_{qi}^B = \beta_q^B$.

If we use the estimator (8.7) based on this model, we define $\hat{\Phi}_{qi}^B$ as follows:

$$\hat{\Phi}_{qi}^B = R_q^B = \frac{\sum_{i=1}^{n_{r,q}}\sum_{k=1}^{M_{qi}^B}w_{qik}}{\sum_{i=1}^{n_q}\sum_{k=1}^{M_{qi}^B}w_{qik}}$$
$$= \frac{\sum_{i=1}^{n_q}\delta_{qi}^B\sum_{k=1}^{M_{qi}^B}w_{qik}}{\sum_{i=1}^{n_q}\sum_{k=1}^{M_{qi}^B}w_{qik}} = \frac{\hat{M}_{r,q}^B}{\hat{M}_q^B} \tag{8.32}$$

where w_{qik} is the estimation weight provided by the GWSM (assuming no non-response) for units *k* in clusters *i* belonging to RHG *q*.

With (8.32), the estimator $\hat{\hat{Y}}^{NRC,B}$ given by (8.31) becomes

$$\begin{aligned}\hat{\hat{Y}}^{NRC,B} &= \sum_{i=1}^{n_r}\sum_{k=1}^{M_i^B}\frac{w_{ik}}{\hat{\Phi}_i^B}y_{ik} = \sum_{q=1}^{Q}\frac{\hat{M}_q^B}{\hat{M}_{r,q}^B}\sum_{i=1}^{n_{r,q}}\sum_{k=1}^{M_{qi}^B} w_{qik}y_{qik} \\ &= \sum_{q=1}^{Q}\frac{\hat{M}_q^B}{\hat{M}_{r,q}^B}\sum_{i=1}^{n_q}\delta_{qi}^B\sum_{k=1}^{M_{qi}^B} w_{qik}y_{qik}.\end{aligned} \tag{8.33}$$

We can look at estimator (8.33) as a ratio estimator in two-phase sampling. Indeed, if the identification of the cluster of Ω^B to be surveyed is considered as the first phase of the sampling design, the "selection" of the subset Ω_r^B of responding clusters makes up the second phase of this design. To obtain the asymptotic bias and the asymptotic variance of $\hat{\hat{Y}}^{NRG,B}$, the Taylor linearisation method is applied, as suggested by Särndal, Swensson and Wretman (1992). Let $\hat{Y}_q^B = \sum_{i=1}^{n_q}\sum_{k=1}^{M_{qi}^B} w_{qik}y_{qik}$. We then get

$$\begin{aligned}\hat{\hat{Y}}^{NRC,B} &\cong \sum_{q=1}^{Q}\left(\hat{Y}_q^B + \sum_{i=1}^{n_q}\delta_{qi}^B\sum_{k=1}^{M_{qi}^B}\frac{w_{qik}}{\beta_q^B}y_{qik} - \frac{\hat{Y}_q^B}{\hat{M}_q^B}\sum_{i=1}^{n_q}\delta_{qi}^B\sum_{k=1}^{M_{qi}^B}\frac{w_{qik}}{\beta_q^B}\right) \\ &= \sum_{q=1}^{Q}\hat{Y}_q^B + \sum_{q=1}^{Q}\left(\sum_{i=1}^{n_q}\frac{\delta_{qi}^B}{\beta_q^B}\sum_{k=1}^{M_{qi}^B} w_{qik}y_{qik} - \frac{\hat{Y}_q^B}{\hat{M}_q^B}\sum_{i=1}^{n_q}\frac{\delta_{qi}^B}{\beta_q^B}\sum_{k=1}^{M_{qi}^B} w_{qik}\right) \\ &= \hat{Y}^B + \sum_{q=1}^{Q}\sum_{i=1}^{n_q}\frac{\delta_{qi}^B}{\beta_q^B}\sum_{k=1}^{M_{qi}^B} w_{qik}\left(y_{qik} - \frac{\hat{Y}_q^B}{\hat{M}_q^B}\right).\end{aligned} \tag{8.34}$$

As in the case of non-response in s^A, we can show that estimator (8.33) is asymptotically unbiased under the conditions of model (8.6). To do this, the expectation of $\hat{\hat{Y}}^{NRC,B}$ from (8.34) is calculated using a conditional approach.

$$
\begin{aligned}
E(\hat{\hat{Y}}^{NRC,B}) &= E_s[(\hat{\hat{Y}}^{NRC,B} \mid \Omega^B)] \\
&\cong E_s\left[\hat{Y}^B + \sum_{q=1}^{Q}\sum_{i=1}^{n_q} \frac{E(\delta_{qi}^B \mid \Omega^B)}{\beta_q^B} \sum_{k=1}^{M_{qi}^B} w_{qik}\left(y_{qik} - \frac{\hat{Y}_q^B}{\hat{M}_q^B}\right)\right] \\
&= E_s\left[\hat{Y}^B + \sum_{q=1}^{Q}\sum_{i=1}^{n_q} \frac{\beta_q^B}{\beta_q^B} \sum_{k=1}^{M_{qi}^B} w_{qik}\left(y_{qik} - \frac{\hat{Y}_q^B}{\hat{M}_q^B}\right)\right] \\
&= E_s\left[\hat{Y}^B + \sum_{q=1}^{Q}\sum_{i=1}^{n_q}\sum_{k=1}^{M_{qi}^B} w_{qik}\left(y_{qik} - \frac{\hat{Y}_q^B}{\hat{M}_q^B}\right)\right] \\
&= E_s\left[\hat{Y}^B + \sum_{q=1}^{Q}\left(\hat{Y}_q^B - \frac{\hat{Y}_q^B}{\hat{M}_q^B}\hat{M}_q^B\right)\right] \\
&= E_s\left[\hat{Y}^B + 0\right] = Y^B.
\end{aligned}
\tag{8.35}
$$

The asymptotic variance of $\hat{\hat{Y}}^{NRC,B}$ is calculated using a conditional approach with the identity

$$Var(\hat{\hat{Y}}^{NRC,B}) = Var_s[E(\hat{\hat{Y}}^{NRC,B} \mid \Omega^B)] + E_s[Var(\hat{\hat{Y}}^{NRC,B} \mid \Omega^B)].$$

From (8.34), we get

$$
\begin{aligned}
Var(\hat{\hat{Y}}^{NRC,B}) \cong & \;{}^{M^A}_{j=1}\;{}^{M^A}_{j'=1} \frac{(\pi_{jj'}^A - \pi_j^A \pi_{j'}^A)}{\pi_j^A \pi_{j'}^A} Z_j Z_{j'} \\
& + E_s \;{}^{Q}_{q=1} \frac{(1-\beta_q^B)}{\beta_q^B} \;{}^{n_q}_{i=1}\;{}^{M_{qi}^B}_{k=1} w_{qik}^2 \; y_{qik} - \frac{\hat{Y}_q^B}{\hat{M}_q^B}\;^2 .
\end{aligned}
\tag{8.36}
$$

The variance (8.36) can be estimated using

$$
\begin{aligned}
\hat{V}ar(\hat{\hat{Y}}^{NRC,B}) = & \sum_{j=1}^{m^A}\sum_{j'=1}^{m^A} \frac{(\pi_{jj'}^A - \pi_j^A \pi_{j'}^A)}{\pi_{jj'}^A \pi_j^A \pi_{j'}^A} Z_j^{NRC} Z_{j'}^{NRC} \\
& + \sum_{q=1}^{Q} \frac{(1-R_q^B)}{(R_q^B)^2} \sum_{i=1}^{n_{r,q}}\sum_{k=1}^{M_{qi}^B} w_{qik}^2 \left(y_{qik} - \frac{\hat{Y}_{r,q}^B}{\hat{M}_{r,q}^B}\right)^2 .
\end{aligned}
\tag{8.37}
$$

If we have auxiliary variables $\mathbf{X}_i^B$ available for all clusters i from Ω^B, the use of the model (8.8) can be considered, with the parameter $\boldsymbol{\beta}^B$ of dimension p^B. With the estimator (8.9) based on this model, we then have

$$\hat{Y}^{NRC,LOGIT,B} = \sum_{i=1}^{n_r}\sum_{k=1}^{M_i^B} \frac{w_{ik}}{\hat{\Phi}_i^{LOGIT,B}} y_{ik} = \sum_{i=1}^{n_r}\sum_{k=1}^{M_i^B} \frac{(1+exp(\hat{\boldsymbol{\beta}}^{T,B}\mathbf{X}_i^B))}{exp(\hat{\boldsymbol{\beta}}^{T,B}\mathbf{X}_i^B)} w_{ik} y_{ik} \tag{8.38}$$

where the estimate $\hat{\boldsymbol{\beta}}^B$ is obtained using the weighted maximum likelihood method with the weights corresponding to the weights w_{ik} from the GWSM.

The estimator $\hat{Y}^{NRC,LOGIT,B}$ is nonlinear and therefore it is not simple to calculate its bias. However, if the estimate $\hat{\Phi}_i^{LOGIT,B}$ turns out to be relatively close to the true response probability Φ_i^B, this bias should be small. Approaches often used in practice at Statistics Canada to estimate the variance of (8.38) are the Jackknife and the Bootstrap methods. If the sampling design used for the selection of the sample s^A is a stratified multi-stage design, the Jackknife estimator for the variance of $\hat{Y}^{NRC,LOGIT,B}$ has the form (6.9).

8.6 TREATMENT OF UNIT NON-RESPONSE

Unit non-response is a kind of total non-response in which one or more units in the cluster, but not all units, did not respond. This type of non-response is particularly important in the context of indirect sampling because it is assumed that all units of the clusters from U^B identified following the selection of the sample s^A are surveyed. If no response is obtained from certain units of the clusters identified to be surveyed, we must then try to correct the situation. In this section, we propose an adjustment to correct unit non-response based on the use of response probabilities.

Following a particular sample design, we again select a sample s^A containing m^A units from population U^A consisting of M^A units. Let $\pi_j^A > 0$ represent the selection probability of unit j. We assume that all m^A units in the sample answered the survey questions.

The target population U^B contains M^B units. This population is divided into N clusters, where cluster i contains M_i^B units. For each unit j in s^A, we identify the units ik of U^B that have a non-zero relationship $l_{j,ik}$ with j, i.e., $l_{j,ik}=1$. For each unit ik identified, it is assumed that a list of the M_i^B units of cluster i containing that unit can be made. Each cluster i represents, by itself, a population U_i^B where $U^B=\bigcup_{i=1}^{N}U_i^B$. Let Ω^B be the set of n clusters identified by the units $j \in s^A$.

In carrying out the survey process, we attempt to survey all units k in clusters i of Ω^B. Unfortunately, for some units in the identified clusters, we are unable to obtain any data. This is unit non-response. This situation is illustrated in Figure 8.3. The arrows indicate that units $j=1$ and $j=2$ from U^A were selected to be part of s^A. Then, clusters $i=1$ and $i=2$ are identified to be surveyed. In cluster $i=1$, only unit 2 responded. In cluster $i=2$, there is no response for unit 3. We assume here that we have a response for at least one unit in each cluster i in Ω^B. Let $s_{r,i}^B$ be the set of responding units in identified cluster i, and let $M_{r,i}^B>0$ be the size of that set.

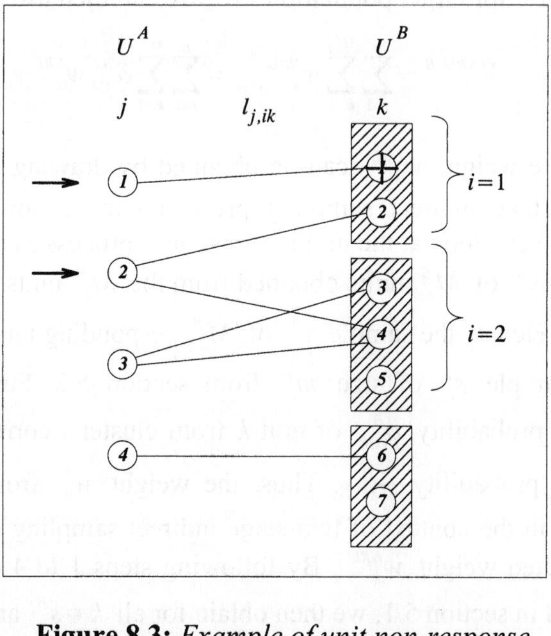

Figure 8.3: *Example of unit non-response*

Let $\delta^B_{(i)k}$ be an indicator variable that takes a value of 1 if unit k of cluster i answers the survey questions, and 0 if not. It is assumed that this variable has a *Bernoulli distribution* with probability $\phi^B_{(i)k}$. In other words, we assume that each unit k in clusters i of U^B has a probability $\phi^B_{(i)k}$ of responding to the survey, i.e.,

$$P(\text{unit } k \in i \text{ reponds} \mid \Omega^{\mathrm{B}}) = P(\delta^B_{(i)k} = 1 \mid \Omega^{\mathrm{B}}) = \phi^B_{(i)k}.$$

In addition, for two units k and k' of a cluster i (or of two different clusters), the indicator variables $\delta^B_{(i)k}$ and $\delta^B_{(i)k'}$ are independent.

For each of the $M^B_{r,i}$ responding units from clusters $i \in \Omega^B$, a variable of interest y is measured. For the target population U^B, we try to estimate the total $Y^B = \sum_{i=1}^N \sum_{k=1}^{M_i^B} y_{ik}$. It is assumed that we have the total number of links L_i^B for each cluster $i \in \Omega^B$.

In applying the GWSM, we want to assign an estimation weight w_{ik}^{NRU} to each responding unit k of cluster i in Ω^B. The superscript "NRU" refers to the non-response of the units. To estimate the total Y^B for target population U^B, we can then use the estimator

$$\hat{Y}^{NRU,B} = \sum_{i=1}^{n} \sum_{k=1}^{M^B_{r,i}} w_{ik}^{NRU} y_{ik} = \sum_{i=1}^{n} \sum_{k=1}^{M^B_i} \delta^B_{(i)k} w_{ik}^{NRU} y_{ik}. \tag{8.39}$$

The weight w_{ik}^{NRU} can be obtained by drawing a parallel with the two-stage indirect sampling presented in section 5.2. In other words, we can look at the unit non-response process as the selection of a sample $s^B_{r,i}$ of $M^B_{r,i}$ units obtained from the M_i^B units in each cluster i of Ω^B. Hence, the sample $s^B_{r,i}$ of $M^B_{r,i}$ responding units corresponds to the sample s_i^B of size m_i^B from section 5.2. Furthermore, the response probability $\phi^B_{(i)k}$ of unit k from cluster i corresponds to the selection probability $\pi^{II}_{(i)k}$. Thus, the weight w_{ik}^{II} from the GWSM obtained in the context of two-stage indirect sampling corresponds to the expected weight w_{ik}^{NRU}. By following steps 1 to 4 of the GWSM presented in section 5.1, we then obtain for all $k \in s^B_{r,i}$ and $i = 1,...,n$:

$$w_{ik}^{NRU} = \frac{w_i}{\phi_{(i)k}^{B}} \tag{8.40}$$

where w_i is given by (2.4).

Theorem 5.1 and Corollaries 5.1 and 5.2 hold in the present context. Following Corollary 5.1, the estimator (8.39) with the weight (8.40) is unbiased.

There exists a fundamental difference however between the theory presented in section 5.2 and that related to unit non-response. This difference lies in the fact that in the context of two-stage indirect sampling, the probability $\pi_{(i)k}^{II}$ is generally known, which unfortunately is not the case for the response probability $\phi_{(i)k}^{B}$. Recall that the estimator $\hat{Y}^{NRU,B}$ is only useful in practice if the value of the response probabilities $\phi_{(i)k}^{B}$ is known for all units k of each set $s_{r,i}^{B}$ of responding units. We are going to try to estimate these probabilities so that we can use the following estimator:

$$\hat{Y}^{NRU,B} = \sum_{i=1}^{n} \sum_{k=1}^{M_{r,i}^{B}} \hat{w}_{ik}^{NRU} y_{ik}, \tag{8.41}$$

where $\hat{w}_{ik}^{NRU} = w_i / \hat{\phi}_{(i)k}^{B}$.

To obtain $\hat{\phi}_{(i)k}^{B}$, we can use one of the following two approaches. The first approach involves considering the set $s_r^B = \bigcup_{i=1}^{n} s_{r,i}^{B}$ of responding units as a whole. Then, the response probabilities $\phi_{(i)k}^{B}$ are estimated without necessarily distinguishing between the different clusters that, as we recall, consist of subpopulations U_i^B from U^B, from which the sets $s_{r,i}^{B}$ are obtained. This approach can be described as *global*. The second approach involves considering each set $s_{r,i}^{B}$ of responding units separately. Then the response probabilities are estimated within each subpopulation U_i^B from U^B.

The two approaches differ from one another at the level of weighting used to estimate the response probabilities $\phi_{(i)k}^{B}$. With the global approach, an estimate can be obtained for the set of the probabilities $\phi^B_{(i)k}$ using the weights w_{ik} from the GWSM. For example, the model (8.6) takes the form: $\phi^B_{(qi)k} = \beta^B_q$. If the estimator (8.7) based on this model is used, the global approach estimator $\hat{\phi}^{GLOB,B}_{(qi)k}$ is then defined as follows:

$$\hat{\phi}^{GLOB,B}_{(qi)k} = R^B_q = \frac{\sum_{i=1}^{n_q}\sum_{k=1}^{M^B_{r,qi}} w_{qik}}{\sum_{i=1}^{n_q}\sum_{k=1}^{M^B_{qi}} w_{qik}} = \frac{\sum_{i=1}^{n_q}\sum_{k=1}^{M^B_{qi}} \delta^B_{(qi)k} w_{qik}}{\sum_{i=1}^{n_q}\sum_{k=1}^{M^B_{qi}} w_{qik}} = \frac{\hat{M}^B_{r,q}}{\hat{M}^B_q} \tag{8.42}$$

where w_{qik} is the estimation weight coming from the GWSM (assuming no non-response) for units k of clusters i belonging to RHG q.

With the *individual approach*, each subpopulation U^B_i is considered individually. Since the GWSM assigns an identical estimation weight to the set of units in each cluster i from Ω^B, the estimation of probabilities $\phi^B_{(i)k}$ for the responding units of each cluster i can be done without weighting. For example, the model (8.6) here takes the form: $\phi^B_{(qi)k} = \beta^B_{qi}$. With this model, we then define $\hat{\phi}^B_{(qi)k}$ as follows:

$$\hat{\phi}^B_{(qi)k} = R^B_{qi} = \frac{M^B_{r,qi}}{M^B_{qi}}, \tag{8.43}$$

where $M^B_{qi} = M^B_i$ and $M^B_{r,qi} = M^B_{r,i}$ for $i \in q$.

In general, the two approaches, global and individual, give different results. In the context of two-stage sampling, direct or indirect, it is nevertheless more natural to consider the clusters (or PSUs) individually, instead of globally. Indeed, because each subpopulation U^B_i is considered as a population itself, the modelling of response probabilities $\phi^B_{(i)k}$ can be performed at the level of each subpopulation U^B_i. Note that this approach is harmonised with the assumption of independence from the second stage of the sampling design mentioned by Särndal, Swensson and Wretman (1992). According to this assumption, the sampling within a PSU (or cluster) must be done independently from the other PSUs. For these reasons, we are going to focus the discussion on the individual approach.

With (8.43), the estimator $\hat{\hat{Y}}^{NRU,B}$ given by (8.41) becomes

$$\begin{aligned}\hat{\hat{Y}}^{NRU,B} &= \sum_{i=1}^{n}\sum_{k=1}^{M_{r,i}^B} \frac{w_i}{\hat{\phi}_{(i)k}^B} y_{ik} \\ &= \sum_{q=1}^{Q}\sum_{i=1}^{n_q} \frac{M_{qi}^B}{M_{r,qi}^B} \sum_{k=1}^{M_{r,qi}^B} w_{qi} y_{qik}.\end{aligned} \tag{8.44}$$

This estimator is nothing more than a ratio estimator within each PSU under two-stage sampling. To obtain the bias and the variance of $\hat{\hat{Y}}^{NRU,B}$, it is useful to prove the following theorem.

Theorem 8.2: Duality of the form of $\hat{\hat{Y}}^{NRU,B}$

Let $\hat{\hat{Y}}_i = \sum_{k=1}^{M_{r,i}^B} y_{ik} / \hat{\phi}_{(i)k}^B$ and $L_i^B = \sum_{k=1}^{M_i^B} L_{ik}^B$. For the clusters $i \in \Omega^B$, we set $\hat{\hat{z}}_{ik} = \hat{\hat{Y}}_i / L_i^B$ for all $k \in U_i^B$. The estimator $\hat{\hat{Y}}^{NRU,B}$ given by (8.41) can then also be written under the form

$$\hat{\hat{Y}}^{NRU,B} = \sum_{j=1}^{M^A} \frac{t_j}{\pi_j^A} \hat{\hat{Z}}_j \tag{8.45}$$

where

$$\hat{\hat{Z}}_j = \sum_{i=1}^{n}\sum_{k=1}^{M_i^B} l_{j,ik} \hat{\hat{z}}_{ik}. \tag{8.46}$$

Proof

The proof of this theorem is the same as that for Theorem 5.1 where, in particular, the selection probability $\pi_{(i)k}^{II}$ is replaced by the estimated response probability $\hat{\phi}_{(i)k}^B$. ■

The estimator $\hat{\hat{Y}}^{NRU,B}$ can therefore be written as a function of units *ik* from U^B, or as a function of units *j* from U^A.

Corollary 8.1: Bias of $\hat{\hat{Y}}^{NRU,B}$

The estimator $\hat{\hat{Y}}^{NRU,B}$ given by (8.44) is asymptotically unbiased for the estimation of Y^B, with respect to the sampling design and under the assumption of the model (8.6).

Proof

The expectation is first decomposed into $E_{\Omega^B}[E(\hat{\hat{Y}}^{NRU,B} \mid \Omega^B)]$ where the first expectation is performed with respect to all possible samples Ω^B from the clusters, and the second expectation is conditional on the clusters of Ω^B. From (8.45) and (8.46), we have

$$E(\hat{\hat{Y}}^{NRU,B} \mid \Omega^B) = \sum_{j=1}^{M^A} \frac{t_j}{\pi_j^A} \sum_{i=1}^{n} \sum_{k=1}^{M^B} l_{j,ik} E(\hat{\hat{z}}_{ik} \mid \Omega^B). \tag{8.47}$$

Now, using (8.43), for $i \in q$, we have

$$\begin{aligned} E(\hat{\hat{z}}_{ik} \mid \Omega^B) &= \frac{1}{L_i^B} E(\hat{\hat{Y}}_i \mid \Omega^B) = \frac{1}{L_{qi}^B} E(\hat{\hat{Y}}_{qi} \mid \Omega^B) \\ &= \frac{1}{L_{qi}^B} E\left(\frac{M_{qi}^B}{M_{r,qi}^B} \sum_{k=1}^{M_{r,qi}^B} y_{qik} \middle| \Omega^B \right) \cong \frac{Y_{qi}}{L_{qi}^B} \end{aligned} \tag{8.48}$$

since the ratio estimator is asymptotically unbiased (Särndal, Swensson and Wretman, 1992). Note that $Y_{qi} = Y_i$ and $L_{qi}^B = L_i^B$ for $i \in q$. Therefore, $E(\hat{\hat{z}}_{ik} \mid \Omega^B) \cong Y_i / L_i^B = z_{ik}$ where z_{ik} is defined in Theorem (4.1).

In fact,

$$E(\hat{\hat{Y}}^{NRU,B} \mid \Omega^B) \cong \sum_{j=1}^{M^A} \frac{t_j}{\pi_j^A} \sum_{i=1}^{n} \sum_{k=1}^{M^B} l_{j,ik} z_{ik}. \tag{8.49}$$

Following the proof of Corollary 5.1, we obtain

$$\begin{aligned} E(\hat{\hat{Y}}^{NRU,B} \mid \Omega^B) &\cong \sum_{j=1}^{M^A} \frac{t_j}{\pi_j^A} \sum_{i=1}^{n} \sum_{k=1}^{M^B} l_{j,ik} z_{ik} \\ &= \sum_{j=1}^{M^A} \frac{t_j}{\pi_j^A} \sum_{i=1}^{N} \sum_{k=1}^{M^B} l_{j,ik} z_{ik} = \hat{Y}^B. \end{aligned} \tag{8.50}$$

Thus, according to Corollary 4.1, $E(\hat{\hat{Y}}^{NRU,B}) \cong Y^B$. ■

Corollary 8.2: Variance of $\hat{\hat{Y}}^{NRU,B}$

The variance formula, with respect to the sampling design, of the estimator $\hat{\hat{Y}}^{NRU,B}$ *coming from* (8.44) *is given by*

$$Var(\hat{\hat{Y}}^{NRU,B}) \cong \sum_{j=1}^{M^A}\sum_{q=1}^{Q}\sum_{i=1}^{N_q}\left(\frac{L_{j,qi}}{L_{qi}^B}\right)^2 \sigma_{qi}^2 + \sum_{j=1}^{M^A}\sum_{j'=1}^{M^A}\frac{(\pi_{jj'}^A - \pi_j^A\pi_{j'}^A)}{\pi_j^A\pi_{j'}^A} Z_j Z_{j'} \tag{8.51}$$

where

$$\sigma_{qi}^2 = \frac{(1-\beta_{qi}^B)}{\beta_{qi}^B}\sum_{k=1}^{M_{qi}^B}\left(y_{qik} - \frac{Y_{qi}}{M_{qi}^B}\right)^2$$

and where β_{qi}^B *is the parameter of the model* (8.6) *in the context of unit non-response.*

Proof

To get a variance formula for $\hat{\hat{Y}}^{NRU,B}$, we start from equation (8.45). As in the proof of Corollary 5.2, we proceed from a conditional argument.

From equation (8.49) and Corollary 4.2, we directly obtain

$$Var_{\Omega^B}\ E(\hat{\hat{Y}}^{NRU,B} \mid \Omega^B) \cong \sum_{j=1}^{M^A}\sum_{j'=1}^{M^A}\frac{(\pi_{jj'}^A - \pi_j^A\pi_{j'}^A)}{\pi_j^A\pi_{j'}^A} Z_j Z_{j'}. \tag{8.52}$$

Now, from (8.45) and (8.46) as well as (8.43), the following result is obtained:

$$\begin{aligned}\hat{\hat{Y}}^{NRU,B} &= \sum_{j=1}^{M^A}\frac{t_j}{\pi_j^A}\sum_{i=1}^{n}\sum_{k=1}^{M^B} l_{j,ik}\frac{\hat{Y}_i^B}{L_i^B} = \sum_{j=1}^{M^A}\frac{t_j}{\pi_j^A}\sum_{i=1}^{n}\frac{L_{j,i}}{L_i^B}\hat{Y}_i \\ &= \sum_{j=1}^{M^A}\frac{t_j}{\pi_j^A}\sum_{i=1}^{n}\frac{L_{j,i}}{L_i^B}\sum_{k=1}^{M_{r,i}^B}\frac{y_{ik}}{\hat{\phi}_{(i)k}^B} \\ &= \sum_{j=1}^{M^A}\frac{t_j}{\pi_j^A}\sum_{q=1}^{Q}\sum_{i=1}^{n_q}\frac{L_{j,qi}}{L_{qi}^B}\frac{M_{qi}^B}{M_{r,qi}^B}\sum_{k=1}^{M_{r,qi}^B} y_{qik}.\end{aligned} \tag{8.53}$$

Then, the conditional variance of $\hat{\hat{Y}}^{NRU,B}$ is calculated to obtain

$$Var(\hat{\hat{Y}}^{NRU,B} \mid \Omega^B) = \sum_{j=1}^{M^A} \frac{t_j}{\pi_j^A} \sum_{q=1}^{Q} \sum_{i=1}^{n_q} \left(\frac{L_{j,qi}}{L_{qi}^B} \right)^2 Var\left(\frac{M_{qi}^B}{M_{r,qi}^B} \sum_{k=1}^{M_{r,qi}^B} y_{qik} \mid \Omega^B \right). \tag{8.54}$$

Since

$$\hat{\hat{Y}}_{qi} = \frac{M_{qi}^B}{M_{r,qi}^B} \sum_{k=1}^{M_{r,qi}^B} y_{qik}$$

is nothing more than a ratio estimator of Y_{qi} for $i \in q$, in the context of a Bernoulli sampling, we have

$$Var(\hat{\hat{Y}}_{qi} \mid \Omega^B) \cong \frac{(1-\beta_{qi}^B)}{\beta_{qi}^B} \sum_{k=1}^{M_{qi}^B} \left(y_{qik} - \frac{Y_{qi}}{M_{qi}^B} \right)^2 = \sigma_{qi}^2. \tag{8.55}$$

From (8.54) and (8.55), and as inspired by the proof of Corollary 5.1, we have

$$\begin{aligned} Var(\hat{\hat{Y}}^{NRU,B} \mid \Omega^B) &\cong \sum_{j=1}^{M^A} \frac{t_j}{\pi_j^A} \sum_{q=1}^{Q} \sum_{i=1}^{n_q} \left(\frac{L_{j,qi}}{L_{qi}^B} \right)^2 \sigma_{qi}^2 \\ &= \sum_{j=1}^{M^A} \frac{t_j}{\pi_j^A} \sum_{q=1}^{Q} \sum_{i=1}^{N_q} \left(\frac{L_{j,qi}}{L_{qi}^B} \right)^2 \sigma_{qi}^2. \end{aligned} \tag{8.56}$$

Finally,

$$E_{\Omega^B}\left[Var(\hat{\hat{Y}}^{NRU,B} \mid \Omega^B) \right] \cong \sum_{j=1}^{M^A} \sum_{q=1}^{Q} \sum_{i=1}^{N_q} \left(\frac{L_{j,qi}}{L_{qi}^B} \right)^2 \sigma_{qi}^2. \tag{8.57}$$

■

The variance (8.51) can be estimated using

$$\hat{V}ar(\hat{\hat{Y}}^{NRU,B}) = \sum_{j=1}^{m^A} \frac{1}{\pi_j^A} \sum_{q=1}^{Q} \sum_{i=1}^{N_q} \left(\frac{L_{j,qi}}{L_{qi}^B} \right)^2 \hat{\sigma}_{qi}^2 + \sum_{j=1}^{m^A} \sum_{j'=1}^{m^A} \frac{(\pi_{jj'}^A - \pi_j^A \pi_{j'}^A)}{\pi_{jj'}^A \pi_j^A \pi_{j'}^A} \hat{\hat{Z}}_j \hat{\hat{Z}}_{j'} \tag{8.58}$$

where
$$\hat{\sigma}_{qi}^2 = \frac{(1-R_{qi}^B)}{(R_{qi}^B)^2} \sum_{k=1}^{M_{r,qi}^B} \left(y_{qik} - \frac{\hat{\hat{Y}}_{qi}}{M_{qi}^B} \right)^2 \tag{8.59}$$

(Cochran, 1977, and Särndal, Swensson and Wretman, 1992).

If we have auxiliary variables $\mathbf{x}_{ik}^B$ available for all units of clusters i from Ω^B, the use of the model (8.8) can be imagined, with parameter $\boldsymbol{\beta}_i^B$ of dimension p^B. With the estimator (8.9) based on this model, we then have

$$\hat{Y}^{NRU,LOGIT,B} = \sum_{i=1}^{n}\sum_{k=1}^{M_{r,i}^B} \frac{w_i}{\hat{\phi}_{(i)k}^{LOGIT,B}} y_{ik} = \sum_{i=1}^{n}\sum_{k=1}^{M_{r,i}^B} \frac{(1+\exp\ (\hat{\boldsymbol{\beta}}_i^{T,B}\ \ \mathbf{x}_{ik}^B))}{\exp\ (\hat{\boldsymbol{\beta}}_i^{T,B}\ \ \mathbf{x}_{ik}^B)} w_i y_{ik} \tag{8.60}$$

where the estimator $\hat{\boldsymbol{\beta}}_i^B$ is obtained using the unweighted maximum likelihood method. If the estimate $\hat{\phi}_{(i)k}^{LOGIT,B}$ turns out to be relatively close to the true response probability $\phi_{(i)k}^B$, the bias of $\hat{Y}^{NRU,LOGIT,B}$ should be small. Approaches often used in practice at Statistics Canada to estimate the variance of (8.60) are the Jackknife and Bootstrap methods. If the sampling design used to select the sample s^A is a multi-stage stratified design, the Jackknife estimator of the variance of $\hat{Y}^{NRU,LOGIT,B}$ has the form (6.9).

8.7 TREATMENT OF ERRORS IN LINKS IDENTIFICATION

The problem of links identification is associated with the situation where it cannot be established if a unit *ik* from U^B is linked to a unit j from U^A. This problem has already been mentioned by Sirken and Nathan (1988) in the context of Network Sampling. More recently, Ardilly and Le Blanc (1999), and Ardilly and Le Blanc (2001), addressed this problem while using the GWSM to weight a survey of homeless persons. Errors in links identification are particularly problematic for the GWSM. Indeed, they can create serious bias problems in the estimates.

As an example, let us consider the case encountered by Ardilly and Le Blanc (2001). Let U^B be the target population of homeless persons, and let U^A represent the set of services (meals, bed, etc.) that are provided to these homeless persons. Using Indirect Sampling, we select a sample s^A of services from U^A, in order to estimate the population U^B of homeless persons. Now, for each service selected in s^A, we are able to identify the homeless person that used this service.

However, the GWSM requires to know all services that the identified homeless person has received, and this is often difficult to get because these persons are usually difficult to interview. This causes errors in the identification of the links.

As always, following a particular sample design, we select a sample s^A containing m^A units from population U^A consisting of M^A units. Suppose that $\pi_j^A > 0$ represents the selection probability of unit *j*. We assume that all m^A units in the sample answered the survey questions.

The target population U^B contains M^B units. This population is divided into *N* clusters, where cluster *i* contains M_i^B units. For each unit *j* in s^A, we identify units *ik* of U^B that have a non-zero relationship $l_{j,ik}$ with *j*, i.e., $l_{j,ik} = 1$. We assume that we can identify **all** relationships $l_{j,ik}$ associated with each unit *j* of s^A. For each identified unit *ik*, we assume that we can make a list of the M_i^B units of cluster *i* containing that unit. Each cluster *i* represents, then, by itself, a population U_i^B where $U^B = \bigcup_{i=1}^{N} U_i^B$. Let Ω^B be the set of *n* clusters identified by units $j \in s^A$.

We survey all units *k* in clusters *i* of Ω^B. Although we can measure the variable of interest *y* for **all** M_i^B units in each cluster *i* of Ω^B, for some units *k*, we fail to determine whether there is a relationship between those units *k* and a unit *j* of U^A. In other words, for **some** units *k* of a cluster $i \in \Omega^B$, it is impossible to determine whether $l_{j,ik} = 1$ or $l_{j,ik} = 0$. Note that, based on interviewing, we know the links $l_{j,ik}$ for all the units *j* from s^A. Hence, we know $l_{j,ik}$ for $j \in s^A$, but we do not know all the $l_{j,ik}$ for $j \in \Omega^{A|B}$ where $\Omega^{A|B} = \{j \in U^A \mid \exists i \in \Omega^B \text{ and } L_{j,i} > 0\}$. The set $\Omega^{A|B}$ contains the units *j* from U^A that have a link to the clusters in Ω^B that were identified at the start by the sample s^A. Let $\dot{\Omega}^{A|B}$ be the set of units from $\Omega^{A|B}$ for which the links have been identified. We can see that s^A is a subset of $\dot{\Omega}^{A|B}$, which is itself in general a subset of $\Omega^{A|B}$. Note that it can happen that some null links $l_{j,ik} = 0$ are identified as being non null, but this seldom happens in practice. Most of the time, some links are missing, which makes $\dot{\Omega}^{A|B}$ a subset of $\Omega^{A|B}$.

The problem of links identification is illustrated in Figure 8.4. The arrows indicate that units $j=1$ and $j=2$ from U^A were selected to be part of s^A. For each of the units $j=1$ and $j=2$, the relationships with the target population U^B can be established. Then, the clusters $i=1$ and $i=2$ are identified to be surveyed. In cluster $i=2$, the relationship between unit 4 of U^B and unit $j=3$ cannot be established, even if the relationship between this unit 4 and unit $j=2$ is known. Still in the cluster $i=2$, the relationship between unit 3 of U^B and unit $j=3$ of U^A cannot be established.

Let $L^B_{r,i}$ be the total number of links identified between cluster i and population U^A. Note that in general $L^B_{r,i} \leq L^B_i$. In addition, because we are assuming that we can identify all **relationships** $l_{j,ik}$ associated with each unit j of s^A, we have $L^B_{r,i} > 0$ for all clusters i in Ω^B.

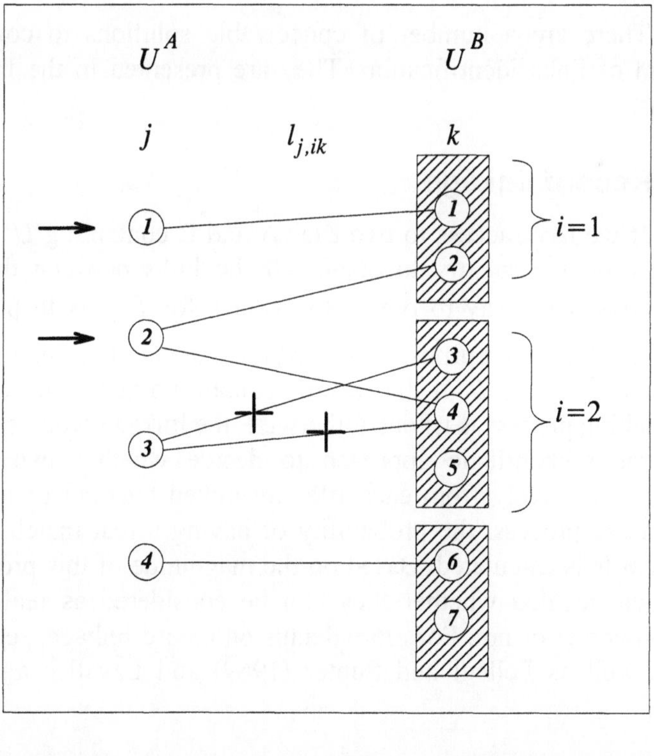

Figure 8.4: *Example of the problem of links identification*

By using only the total number $L^B_{r,i}$ of links found, we overestimate the total Y^B. This can be seen from the expression of $\hat{Y}^B$ given by Corollary 4.3. Indeed, if L^B_i is replaced by $L^B_{r,i}$ in (4.16), we obtain

$$\hat{Y}^{NRL,B} = \sum_{j=1}^{M^A} \frac{t_j}{\pi_j^A} \sum_{i=1}^{N} Y_i \frac{L_{j,i}}{L^B_{r,i}} \qquad (8.61)$$

where the superscript "NRL" stands for non-response within the links. Since $L^B_{r,i} \leq L^B_i$, we have directly $\hat{Y}^{NRL,B} \geq \hat{Y}^B$. Likewise, $E(\hat{Y}^{NRL,B}) \geq E(\hat{Y}^B) = Y^B$ and thus $\hat{Y}^{NRL,B}$ is a biased estimator of Y^B.

It is important to note that there is no problem in obtaining the quantity $L_{j,i}$ in (8.61) because it is assumed that **all** the relationships $l_{j,ik}$ associated to each unit j of s^A can be identified. Looking at (4.16) (or (8.61)), it is seen that this quantity must be known only for the units j of s^A.

There are a number of conceivable solutions to correct the problem of links identification. They are presented in the following sections.

8.7.1 Record linkage

If we have access to two files A and B containing U^A and U^B, respectively, we can try to obtain all the links between these two populations. One way to obtain the values for $l_{j,ik}$ is to perform a *record linkage*. The purpose of record linkage is to link the records of the two files *A* and *B*. If the records contain unique identifiers, then the matching process is trivial. Otherwise, the linkage process needs to use some probabilistic approach to decide whether two records, coming respectively from each file, are linked together or not. With this linkage process, the probability of having a real match between two records is calculated. Based on the magnitude of this probability, it is then decided whether they can be considered as really being linked together or not. For more details on record linkage, see section 9.1, as well as Fellegi and Sunter (1969) and Lavallée and Caron (2001).

If obtaining the values $l_{j,ik}$ reveals to be too difficult because, for example, of the size of the files A and B, one can restrict the record linkage to the units k from the clusters i in Ω^B and the population U^A. This is sufficient because, as mentioned earlier, we already know the $l_{j,ik}$ for $j \in s^A$, but we do not know all the $l_{j,ik}$ for the set $\Omega^{A|B}$ containing the units j from U^A that have a link to the clusters in Ω^B.

One can also use record linkage to try evaluating $L_{j,i}$ between the clusters i in Ω^B and the population U^A. As we can see from (8.61), it is sufficient to obtain the quantities $L_{j,i}$, rather than the individual links $l_{j,ik}$, for using the estimator $\hat{Y}^B$.

8.7.2 Modelling

It is possible to estimate the probabilities $\phi_{j,ik}$ of a link between the units j and ik by using a logistic-type model with vectors $\mathbf{x}_j^A$ and $\mathbf{x}_{ik}^B$ of auxiliary variables. Recall that $\dot{\Omega}^{A|B}$ is the set of units from $\Omega^{A|B}$ for which the links have been identified. With the estimated probabilities $\hat{\phi}_{j,ik}$, we can produce estimates $\hat{l}_{j,ik} = \hat{\phi}_{j,ik}$ for the units j contained in the set $\Omega^{A|B} \setminus \dot{\Omega}^{A|B}$. This can be seen as imputing links $l_{j,ik}$ for these units (see Ardilly and Le Blanc, 2001). Note that it is important to make maximum use of all constraints and information that would be associated with the values for $l_{j,ik}$ during modelling.

With $\hat{l}_{j,ik}$, we obtain

$$\hat{L}_{ik}^B = \sum_{j \in \dot{\Omega}^{A|B}} l_{j,ik} + \sum_{j \in \Omega^{A|B} \setminus \dot{\Omega}^{A|B}} \hat{l}_{j,ik} \tag{8.62}$$

and then

$$\hat{L}_i^B = \sum_{k=1}^{M_i^B} \hat{L}_{ik}^B . \tag{8.63}$$

It is also possible to concentrate on L_i^B as a whole, without reference to the population U^A. If we have auxiliary variables $\mathbf{X}_i^B$ available for all clusters i from Ω^B, the approach is then to use a *log-linear model* of the type $\log(L_i^B) = \boldsymbol{\beta}^{T,B}\mathbf{X}_i^B$, where $\boldsymbol{\beta}^B$ is a column vector of parameters of size p^B. The estimate $\hat{\boldsymbol{\beta}}^B$ of $\boldsymbol{\beta}^B$ can be obtained using the unweighted maximum likelihood method (see Bishop, Finberg and Holland, 1975). With $\hat{\boldsymbol{\beta}}^B$, we compute

$$\hat{L}_i^{LGLIN,B} = \exp(\hat{\boldsymbol{\beta}}^{T,B}\mathbf{X}_i^B) \tag{8.64}$$

Using (8.63) (or (8.64)), we then construct the estimator

$$\hat{\hat{Y}}^{NRL,B} = \sum_{j=1}^{M^A} \frac{t_j}{\pi_j^A} \sum_{i=1}^{N} Y_i \frac{L_{j,i}}{\hat{L}_i^B}. \tag{8.65}$$

It is possible to calculate the asymptotic bias of (8.65) by using Taylor linearisation. The linearised estimator is then given by

$$\hat{\hat{Y}}^{NRL,B} \cong \sum_{j=1}^{M^A} \frac{t_j}{\pi_j^A} \sum_{i=1}^{N} Y_i \frac{L_{j,i}}{(L_i^B)^2}(2L_i^B - \hat{L}_i^B). \tag{8.66}$$

Let $\boldsymbol{E}_s(\cdot)$ denote the expected value carried out in relation to all possible samples of s^A. To calculate the asymptotic bias of the estimator $\hat{\hat{Y}}^{NRL,B}$, we proceed as follows from (8.66):

$$\begin{aligned} E(\hat{\hat{Y}}^{NRL,B}) &= E_s[E(\hat{\hat{Y}}^{NRL,B} \mid s^A)] \\ &\cong E_s\left[\sum_{j=1}^{M^A} \frac{t_j}{\pi_j^A} \sum_{i=1}^{N} Y_i \frac{L_{j,i}}{(L_i^B)^2}(2L_i^B - E(\hat{L}_i^B \mid s^A))\right]. \end{aligned} \tag{8.67}$$

If $E(\hat{L}_i^B \mid s^A) = L_i^B$, we then have

$$\begin{aligned} E(\hat{\hat{Y}}^{NRL,B}) &\cong E_s\left[\sum_{j=1}^{M^A} \frac{t_j}{\pi_j^A} \sum_{i=1}^{N} Y_i \frac{L_{j,i}}{(L_i^B)^2}(2L_i^B - L_i^B)\right] \\ &= E_s\left[\sum_{j=1}^{M^A} \frac{t_j}{\pi_j^A} \sum_{i=1}^{N} Y_i \frac{L_{j,i}}{(L_i^B)^2} L_i^B\right] = E_s\left[\hat{Y}^B\right] = Y^B. \end{aligned} \tag{8.68}$$

The last line follows directly from Corollary 4.1. The unbiased nature of the estimator $\hat{\hat{Y}}^{NRL,B}$ depends, however, on the unbiased nature of the estimator of L_i^B. In practice, it is not easy to obtain an unbiased estimator of L_i^B.

For the survey of homeless persons, Ardilly and Le Blanc (2001) suggested making an assumption of regularity to impute some relationships $l_{j,ik}$ to 1, which is actually the same as modelling the quantity L_i^B. Ardilly and Le Blanc (2001), however, have questioned if the assumption of regularity may not be satisfied in practice.

Approaches used at Statistics Canada to estimate the variance of estimators such as (8.65) are the Jackknife and the Bootstrap methods. If the sampling design used for the selection of the sample s^A is a stratified multi-stage design, the Jackknife estimator for the variance of (8.65) has the form (6.9).

8.7.3 Estimating the proportion of links

Another way of solving the problem of links identification is to concentrate on the quantity $\tilde{\theta}_{j,i} = L_{j,i} / L_i^B$, rather than on the number of links $L_{j,i}$. In order to make the estimator (4.16) unbiased, we need only to ensure that $\sum_{j=1}^{M^A} \tilde{\theta}_{j,i} = 1$ (see Ernst, 1989, as well as Lavallée and Deville, 2002). Thus, it is not necessary to know all values of $L_{j,i}$ for $i \in \Omega^{A|B}$, but simply the values $\tilde{\theta}_{j,i}$ for $i \in s^A$, making sure that $\sum_{j=1}^{M^A} \tilde{\theta}_{j,i} = 1$.

It is important to note that $\tilde{\theta}_{j,i}$ can be defined in a general way, without reference to the links $L_{j,i}$. As in section 4.5, some $\tilde{\theta}_{j,i}$ can be defined arbitrarily by keeping $\sum_{j=1}^{M^A} \tilde{\theta}_{j,i} = 1$, which means that we can also use the unbiased estimator:

$$\tilde{Y}^B = \sum_{j=1}^{M^A} \frac{t_j^A}{\pi_j^A} \sum_{i=1}^{N} \tilde{\theta}_{j,i} Y_i \tag{8.69}$$

Of course, the precision of the estimator (8.69) will be subject to the choice of the values $\tilde{\theta}_{j,i}$. As an application, this approach was used by Bankier (1983) to produce statistics from tax data.

8.7.4 Calibration

Another possible solution to correct the overestimation of the estimator $\hat{Y}^{NRL,B}$ given by (8.61) is calibration. If we have auxiliary variables $\mathbf{x}_{ik}^{B}$ correlated with the variable of interest y_{ik}, it is indeed possible to correct, at least in part, the overestimation of the estimator $\hat{Y}^{NRL,B}$. By calibrating the estimator $\hat{Y}^{NRL,B}$ on the known total $\mathbf{X}^{B} = \sum_{i=1}^{N} \sum_{k=1}^{M_i^B} \mathbf{x}_{ik}^{B}$, since this known total is correlated with the total Y^B, a part of the overestimation of $\hat{Y}^{NRL,B}$ will be corrected. Note that the more the variables $\mathbf{x}_{ik}^{B}$ and y_{ik} are correlated, the more efficient the correction of the overestimation will be.

Let $Z_j^{NRL} = \sum_{i=1}^{N} Y_i L_{j,i} / L_{r,i}^{B}$. The equation (8.61) then becomes:

$$\hat{Y}^{NRL} = \sum_{j=1}^{M^A} \frac{t_j}{\pi_j^A} Z_j^{NRL}. \tag{8.70}$$

The calibration estimator $\hat{Y}^{NRL,CAL,B} = \sum_{j=1}^{m^A} w_j^{CAL,A} Z_j^{NRL}$ associated with the GWSM in the presence of non-response within the links is determined from the formulation (7.48). An estimator of the form (7.50) is obtained where the variable Z_j is replaced by Z_j^{NRL}.

While calibration offers an attractive solution to the problem of links identification, it depends on the availability of auxiliary variables $\mathbf{x}_{ik}^{B}$ correlated with the variable of interest y_{ik}, which is not always the case in practice. The best solution is still to measure L_i^B exactly or, failing that, to obtain an estimate $\hat{L}_i^B$ that is as close as possible to L_i^B.

8.7.5 Proportional adjustments

Xu and Lavallée (2006) proposed to solve the problem of links identification by directly estimating L_i^B using a *proportional adjustment*.

Let $\Omega_i^{A|B} = \{j \in U^A \mid i \in \Omega^B \text{ and } L_{j,i} > 0\}$ and let $M_i^{A|B}$ be the number of units j in $\Omega_i^{A|B}$. The set $\Omega_i^{A|B}$ contains the units from U^A that have a link to the cluster i in Ω^B. Note that, in general, we have $\Omega_i^{A|B} \cap \Omega_{i'}^{A|B} \neq \emptyset$, and thus, $\sum_{i=1}^{N} M_i^{A|B} \geq M^A$. Since for a given cluster i in Ω^B, $L_{j,i}$ are non null only for the units j of the set $\Omega_i^{A|B}$, we directly obtain $L_i^B = \sum_{j=1}^{M^A} L_{j,i} = \sum_{j=1}^{M_i^{A|B}} L_{j,i}$. Finally, we have $M_i^{A|B} \leq L_i^B$.

The set $\Omega_i^{A|B}$ contains the units j from U^A that have a link to the cluster i, whether they are in the sample s^A or not. Let us define $s_i^{A|B} = \{j \in s^A \mid i \in \Omega^B \text{ et } L_{j,i} > 0\}$ and let $m_i^{A|B}$ be the number of units j in $s_i^{A|B}$. The set $s_i^{A|B}$ contains the units from s^A that are linked to the cluster i. We can see $s_i^{A|B}$ as a "sample" of $\Omega_i^{A|B}$. Let the "selection probability" be $\pi_{j|i}^{A|B} = P(j \in s_i^{A|B} \mid j \in \Omega_i^{A|B})$. It should be noted that $\pi_{j|i}^{A|B}$ is a function of π_j^A. Accordingly, we can define the following estimator for L_i^B:

$$\hat{L}_i^{PROP,B} = \sum_{j=1}^{m_i^{A|B}} \frac{t_{j|i}^{A|B}}{\pi_{j|i}^{A|B}} L_{j,i} \tag{8.71}$$

where $t_{j|i}^{A|B} = 1$ if $j \in s_i^{A|B}$, and 0 otherwise. It is clear that $E(\hat{L}_i^{PROP,B}) = \sum_{j=1}^{M_i^{A|B}} L_{j,i} = L_i^B$ and thus, the estimator (4.16) used with (8.71) is asymptotically unbiased for the estimation of Y^B.

One of the difficulties in using the estimator (8.71) involves calculating the probabilities $\pi_{j|i}^{A|B}$. If the π_j^A are relatively homogenous, then we can use $\hat{\pi}_{j|i}^{A|B} = m_i^{A|B} / M_i^{A|B} = f_i^{A|B}$. This approach allows us to focus not on the links themselves and the quantities $L_{j,i}$ and L_i^B, but solely on the units j from U^A that are involved in the survey of the units i in Ω^B.

Unfortunately, $M_i^{A|B}$ is often unavailable because of the error in observing the links, which makes it difficult to use the estimator (8.71). In this case, we can try to estimate L_i^B by *global proportional adjustment.* For this adjustment, the variations between the clusters i from U^B are ignored. Thus, s^A is considered to be a "sample" of $\Omega^{A|B}$ and the "selection probability" is defined as $\pi_j^{A|B} = P(j \in s^A \mid j \in \Omega^{A|B})$. Note that $\pi_j^{A|B}$ is a function of π_j^A. Accordingly, in order to estimate L_i^B, we can use

$$\hat{L}_i^{GPROP,B} = \sum_{j=1}^{m_i^{A|B}} \frac{t_j^{A|B}}{\pi_j^{A|B}} L_{j,i} \tag{8.72}$$

As for the estimator (8.71), one of the difficulties in using the estimator (8.72) is obtaining $\pi_j^{A|B}$. Here, we can try to use the approximation $\hat{\pi}_j^{A|B} = m^A / M^{A|B} = f^{A|B}$. In practice, $M^{A|B}$ (or $f^{A|B}$) may be easier to obtain than $M_i^{A|B}$ (or $f_i^{A|B}$).

It is advisable to make maximum use of all constraints and information that would help to calculate the values of $M_i^{A|B}$ or $M^{A|B}$. For example, in the context of longitudinal surveys of individuals within households, one can rely on the fact that the household composition is often relatively stable through time. As in section 6.3, U^A is the population of individuals at the starting wave, and U^B is the target population of individuals within households at a later wave. Letting the clusters i correspond to the households, and assuming that the household composition is relatively stable through time, we can then assume that $M_i^{A|B} \approx M_i^B$. For further details, see Xu and Lavallée (2006).

In the case where some links between s^A and U^B (or between Ω^B and U^A) are unknown or incorrectly identified during the survey process, it can be appropriate to select a subsample from s^A, and to conduct an assessment of the links for this subsample.

The method is as follows: We select a subsample s'^A of m'^A units from s^A, according to some sample design, in order to get the exact links for this subsample. Assuming that the subsample s'^A leads us at the end to $M'^{A|B}$ units from U^A, we can estimate the selection probability $\pi_j^{A/B} = P(j \in s^A \mid j \in \Omega^{A/B})$ for the units j of the sample s^A, using $\hat{\pi}_j^{A/B} = m'^A / M'^{A|B}$. We then compute the estimate

$$\hat{L}_i^{SUBS,B} = \frac{M'^{A/B}}{m'^A} \sum_{j=1}^{m^A} L_{j,i} \tag{8.73}$$

In general, the different proportional adjustments proposed in this section offer good alternatives to decrease the bias of (8.61) due to the problem of links identification. Using data from SLID, Xu and Lavallée (2006) found that all the proposed proportional adjustments perform well both for reducing the bias and the variance of cross-sectional estimates of totals. Hurand (2006) obtained the same results with agriculture data similar to the ones described in sections 7.4.4 and 9.3.1. Although all methods were performing well, Hurand (2006) found that the method based on subsampling provided the best results.

CHAPTER 9

GWSM AND RECORD LINKAGE

Data from different sources are increasingly being combined to augment the amount of information that we have. Often, the databases are combined using record linkage. When the files involved have a unique identifier that can be used, the linkage is done directly using the identifier as a matching key. When there is no unique identifier, a *probabilistic linkage* is used. In that case, a record on the first file is linked to a record from the second file with a certain probability. Then, a decision is made on whether this link is a true link or not. Note that this process usually requires a certain amount of manual resolution.

We again consider the production of an estimate of a total of one target clustered population U^B when using a sample s^A selected from another population U^A that is linked to the first population. However, we assume that the two populations have been linked together using probabilistic record linkage. Note that this type of linkage often leads to a complex linkage between the two populations.

In this chapter, we will try to answer the following questions:

a) Can we use the GWSM to handle the estimation problems related to populations linked together through record linkage?

b) Can we adapt the GWSM to take into account the linkage weights issued from record linkage?

c) Can the GWSM help in reducing the manual resolution required by record linkage?

d) If there is more than one approach to use the GWSM, is there a "better" approach?

It will be seen that the answer is clearly yes to (a) and (b). However, for question (c), it will be shown that there is unfortunately a price to pay in terms of an increase to the sample size, and therefore to the collection costs. For question (d), although there is no definite answer, some approaches seem to generally be more appropriate.

9.1 RECORD LINKAGE

The concepts of *record linkage* were introduced by Newcome *et al.*. (1959), and formalised in the mathematical model of Fellegi and Sunter (1969). As described by Bartlett *et al.* (1993), *record linkage* is the process of bringing together two or more separately recorded pieces of information pertaining to the same unit (individual or business). Record linkage is sometimes called *exact matching*, in contrast to *statistical matching*. This last process attempts to link files that have few units in common. In this case, linkages are based on similar characteristics rather than unique identifying information. To learn more about statistical matching, see Budd and Radner (1969), Budd (1971), Okner (1972) and Singh *et al.* (1993). In this chapter, we will restrict ourselves to the context of record linkage. However, the theory presented can also be used for statistical matching.

Suppose that we have two files A and B containing characteristics respectively relating to two populations U^A and U^B. The two populations are related in a way. They can represent, for example, exactly the same population, where each of the files contains a different set of characteristics of the units of that population. They can also represent different populations, but naturally linked to one another. For example, one population can be one of parents, and the other population one of children belonging to the parents, as illustrated in Figure 1.2. Note that the children usually live in households that can be viewed as clusters.

Another example is one of the creation of Statistics Canada's Whole Farm Database. This example was presented before in section 7.4.4. The first population is a list of farms from the Canadian Census of Agriculture, and the second population is a list of taxation records (or income tax reports) from the Canada Revenue Agency (CRA). In the first population, each farm is identified by a unique identifier called the FarmID and some additional variables such as the name and address of the farm operators that are obtained from the Census questionnaire. The second population consists of tax reports of individuals having declared some form of agricultural income. These individuals live in households (or clusters). The unique identifier on those records is a corporation number or a social insurance number, depending on whether or not the business is incorporated. Note that each income tax report submitted to CRA contains similar variables (name and address of respondent, etc.) as those obtained by the Census of Agriculture.

The purpose of record linkage is to link the records of the two files A and B. If the records contain unique identifiers, then the matching process is trivial. Unfortunately, often a unique identifier is not available and then the linkage process needs to use some probabilistic approach to decide whether two records, coming respectively from each file, are linked together or not. With this linkage process, the probability of having a real match between two records is calculated. Based on the magnitude of this probability, it is then decided whether they can be considered as really being linked together or not.

Formally, we consider the product space $A \times B$ from the two files A and B. Let j indicate a record (or unit) from file A (or population U^A) and k a record (or unit) from file B (or population U^B). For each pair (j,k) of $A \times B$, we compute a *linkage weight* reflecting the degree to which the pair (j,k) has a true link. The higher the linkage weight is, the more likely the pair (j,k) has a true link. The linkage weight is commonly based on the ratio of the conditional probabilities of having a match υ and an unmatch $\bar{\upsilon}$, given the result of the outcome of the comparison $\Delta_{\varsigma jk}$ of the characteristic ς of record j from A and k from B, $\varsigma = 1, \ldots, p$. Thus, the linkage weight can be defined as follows:

$$\begin{aligned}\dot{\theta}_{jk} &= \log_2 \left\{ \frac{P(\upsilon_{jk} \mid \Delta_{1jk}\Delta_{2jk}\ldots\Delta_{pjk})}{P(\bar{\upsilon}_{jk} \mid \Delta_{1jk}\Delta_{2jk}\ldots\Delta_{pjk})} \right\} \\ &= \dot{\theta}_{1jk} + \dot{\theta}_{2jk} + \ldots + \dot{\theta}_{pjk} + \dot{\theta}_{\bullet jk}\end{aligned} \tag{9.1}$$

where $\dot{\theta}_{\varsigma jk} = \log_2 \left\{ \frac{P(\Delta_{\varsigma jk} \mid \upsilon_{jk})}{P(\Delta_{\varsigma jk} \mid \bar{\upsilon}_{jk})} \right\}$ for $\varsigma = 1, \ldots, p$, and $\dot{\theta}_{\bullet jk} = \log_2 \left\{ \frac{P(\upsilon_{jk})}{P(\bar{\upsilon}_{jk})} \right\}$.

The mathematical model proposed by Fellegi and Sunter (1969) considers the probabilities of an error in the linkage of units j from A and k from B. The linkage weight is then defined as

$$\theta_{jk}^{FS} = \sum_{\varsigma=1}^{p} \theta_{\varsigma jk}^{FS},$$

where $\theta^{FS}_{\varsigma jk} = \log_2(\varphi_{\varsigma jk}/\overline{\varphi}_{\varsigma jk})$ if characteristic ς of pair *(j,k)* is linked, and $\theta^{FS}_{\varsigma jk} = \log_2((1-\varphi_{\varsigma jk})/(1-\overline{\varphi}_{\varsigma jk}))$ otherwise. The expressions used here are $\varphi_{\varsigma jk} = P(\Delta_{\varsigma jk} | \upsilon_{jk})$ and $\overline{\varphi}_{\varsigma jk} = P(\Delta_{\varsigma jk} | \overline{\upsilon}_{jk})$. Moreover, it is assumed that the p comparisons are independent.

The linkage weights given by (9.1) are defined on the set $\Re$ of real numbers, i.e., $\dot{\theta}_{jk} \in]-\infty,+\infty[$. When the ratio of the conditional probabilities of having a match υ_{jk} and an unmatch $\overline{\upsilon}_{jk}$ is equal to 1, we get $\dot{\theta}_{jk} = 0$. When this ratio is close to 0, $\dot{\theta}_{jk}$ approaches $-\infty$. It can however be practical to define the linkage weights on $[0,+\infty[$. This can be achieved by taking the antilogarithm of $\dot{\theta}_{jk}$. We then obtain the following linkage weight θ_{jk}:

$$\theta_{jk} = \frac{P(\upsilon_{jk} \mid \Delta_{1jk}\Delta_{2jk}...\Delta_{pjk})}{P(\overline{\upsilon}_{jk} \mid \Delta_{1jk}\Delta_{2jk}...\Delta_{pjk})}. \tag{9.2}$$

Note that the linkage weight θ_{jk} is equal to 0 when the conditional probabilities of having a match υ_{jk} are equal to 0. In other words, we have $\theta_{jk} = 0$ when the probability of having a true link for *(j,ik)* is zero.

Once a linkage weight θ_{jk} has been computed for each pair *(j,k)* of $A \times B$, we need to decide whether the linkage weight is sufficiently large to consider the pair *(j,k)* as being linked. For this, a *decision rule* is generally used. With the approach of Fellegi and Sunter (1969), we choose an upper threshold θ_{High} and a lower threshold θ_{Low} to which each linkage weight θ_{jk} is compared. The decision is made as follows:

$$D(j,k) = \begin{cases} \text{link} & \text{if } \theta_{jk} \geq \theta_{High} \\ \text{possible link} & \text{if } \theta_{Low} < \theta_{jk} < \theta_{High} \\ \text{non-link} & \text{if } \theta_{jk} \leq \theta_{Low}. \end{cases} \tag{9.3}$$

The lower and upper thresholds θ_{Low} and θ_{High} are determined by error bounds that are determined prior to the record linkage process, based on false links and false non-links. When applying the decision rule (9.3), a manual resolution is necessary to make a decision concerning the pairs whose linkage weights are between the lower and upper thresholds. This is generally done by looking at the data, and also by using auxiliary information. In the agriculture example, variables such as date of birth, address and postal code, which are available on both files, are used for this purpose. The application of decision rule (9.3) leads to the definition of an indicator variable l_{jk} such that $l_{jk} = 1$ if the pair *(j,k)* is considered to be a link, and 0 otherwise. Note that the decision rule (9.3) does not prevent the existence of complex links such as those illustrated in Figure 2.1.

By using an automated system and by applying a probabilistic method, the record linkage process can contain some errors. This problem has been discussed in several papers, namely Bartlett *et al.* (1993), Belin (1993) and Winkler (1995). Linkage errors are out of the scope of this book, and thus will only be briefly covered in certain occasions in this chapter.

9.2 GWSM ASSOCIATED WITH RECORD LINKAGE

Let U^A be the population containing M^A units and U^B be the population consisting of *N* clusters where each cluster *i* contains M_i^B units. With record linkage, linked are established between the populations U^A and U^B using a probabilistic process. As mentioned previously, record linkage uses a decision rule *D* such as the one given by (9.3) to decide whether or not there is a link between unit *j* from U^A and unit *ik* from U^B. Once the links are established, we then have two populations U^A and U^B linked together and where the links are identified by the indicator variable $l_{j,ik}$. Recall that the decision rule (9.3) does not prevent complex links from being obtained.

Although the links can be complex, the GWSM can be used to estimate the total Y^B from population U^B using a sample s^A obtained from population U^A. Therefore, the answer is yes to question (a) expressed at the beginning of this chapter. The GWSM used with populations U^A and U^B linked together by record linkage with decision rule (9.3) will be called, in the rest of the chapter, the *classical approach*.

It should be noted that these estimates obtained by the application of the GWSM can be biased if Constraint 2.1 presented in chapter 2 is not satisfied. In this case, estimator (2.1) underestimates the total Y^B. To resolve this problem, a practical solution is to group two clusters so that at least one non-zero link $l_{j,ik}$ is obtained for each cluster *i*. This solution generally requires manual resolution. Another solution is to create, or impute, a link by randomly choosing a link within the cluster. The link with the largest linkage weight $\theta_{j,ik}$ can also be chosen. Note that for a unit j from U^A, there may only be links $l_{j,ik} = 0$ with all units ik from U^B. However, this is not a problem since we are only interested in the coverage of the target population U^B, and not the one of U^A.

Now, with the classical approach, the use of the GWSM is based on links identified by the indicator variable $l_{j,ik}$. Is it necessary to establish whether or not there is positively a link for each pair *(j,ik)*? Would it be easier to use the linkage weights $\theta_{j,ik}$ (without decision rules) to estimate the total Y^B? These questions lead to question (b), that is, if it is possible to adapt the GWSM to take into account the linkage weights issued from record linkage. The answer to this question is yes, as it was shown in section 4.5 that it was possible to extend the use of the GWSM to weighted links.

Recall that by presenting the WSM in the context of longitudinal surveys, Ernst (1989) proposed the use of constants α in the definition of estimation weights. Setting $\widetilde{\theta}_{j,ik} = \theta_{j,ik} / \theta_i^B$, where $\theta_i^B = \sum_{j=1}^{M^A} \sum_{k=1}^{M_i^B} \theta_{,ik}$, a version of the GWSM was obtained in section 4.5, constructed from these constants. Coming back to the context of longitudinal surveys, we saw in section 3.3 that Kalton and Brick (1995) looked at the determination of optimal values for the constants α of Ernst (1989) by looking to minimise the variance. They concluded that: "in the two-household case, the equal household weighting scheme minimises the variance of the household weights around the inverse selection probability weight when the initial sample is an epsem[1] one". They also added that: "in the case of an approximately epsem sample, the equal household weighting scheme should be close to the optimal, at least for the case where the members of the household at time t come from one or two households at the initial wave". Recall that if s^A is a sample of persons, considering the fact that the persons represent households of size 1, the equal weighting of households and the equal weighting of persons are equivalent, which corresponds to the fair share method describe in section 3.2. This suggests that, for the version of the GWSM described in section 4.5, we should be close to the optimal values by setting the values of the constants α to zero for some units and to an

---

[1] "epsem" stands for equal probability selection method.

equal positive value for all other units of the cluster. As with $\alpha_{j,ik} = l_{j,ik} / L_i^B$, the desired types of values are directly obtained, and the classical approach should then produce variances close to the minimum for the estimate of the total Y^B. This result was proved in a formal way in section 4.6.3 for the case of simple random sampling.

In the present section, three different approaches are given where the GWSM uses the linkage weights $\theta_{j,ik}$. The first approach is to use all the non-zero links (i.e., with $\theta_{j,ik} > 0$) identified through the record linkage process with their respective linkage weights. The second approach is the one where we use all the non-zero links with linkage weights above a given threshold θ_{High}. The third approach consists of randomly choosing the links proportionally to $\theta_{j,ik}$.

9.2.1 Approach 1: use all non-zero links with their respective linkage weights

With the use of all non-zero links with the GWSM, it can be justified to give more importance to the links that have a larger linkage weight $\theta_{j,ik}$, compared to those that have a small linkage weight. By definition, for each pair *(j,ik)* obtained from crossing populations U^A and U^B, the linkage weight $\theta_{j,ik}$ reflects the tendency of the pair *(j,ik)* to have a true link. In this case, instead of using the indicator variable $l_{j,ik}$ identifying whether or not there is a link between unit *j* from U^A and unit *k* of cluster *i* from U^B, we can use the linkage weight $\theta_{j,ik}$ obtained in the first steps of the record linkage process. Note that this implies the elimination of the manual resolution since no decision rule is used.

The application of this approach assumes, of course, that the file with the linkage weights is available. In practice, the only file available is often the final file, once the linkage process ends, after manual resolution. In this case, the linkage weights are not generally available (only the indicator variables $l_{j,ik}$ remain) and the three proposed approaches are then no longer pertinent.

For each unit *j* selected in s^A, we identify the units *ik* of U^B that have a non-zero linkage weight with unit *j*, i.e., $\theta_{j,ik} > 0$. Let $\Omega^{RL,B} = \{i \in U^B \mid \exists j \in s^A \text{ and } \theta_{j,i} > 0\}$ with $\theta_{j,i} = \sum_{k=1}^{M_i^B} \theta_{j,ik}$ be the set of n^{RL} clusters identified by the units $j \in s^A$, where "RL" stands for record linkage. Note that because we use all linkage weights greater than zero, we have $n^{RL} \geq n$, where n is the number of clusters identified by the classical approach.

To estimate the total Y^B of the population U^B, one can use the estimator

$$\hat{Y}^{RL,B} = \sum_{i=1}^{n^{RL}} \sum_{k=1}^{M_i^B} w_{ik}^{RL} y_{ik} \tag{9.4}$$

where w_{ik}^{RL} is the estimation weight obtained from the GWSM. This weight is obtained by directly replacing the indicator variable $l_{j,ik}$ with the linkage weight $\theta_{j,ik}$ in the steps of the GWSM described in chapter 2. The following steps are then obtained.

Steps of the GWSM for approach 1

Step 1: For each unit k of the clusters i from $\Omega^{RL,B}$, the initial weight $w_{ik}'^{RL}$ is calculated, that is:

$$w_{ik}'^{RL} = \sum_{j=1}^{M^A} \theta_{j,ik} \frac{t_j}{\pi_j^A} \tag{9.5}$$

where $t_j = 1$ if $j \in s^A$, and 0 otherwise. Note that a unit ik having no link with any unit j from U^A automatically has an initial weight of zero.

Step 2: For each unit k of the clusters i from $\Omega^{RL,B}$, we calculate

$$\theta_{ik}^B = \sum_{j=1}^{M^A} \theta_{j,ik} .$$

Step 3: The final weight w_i^{RL} is calculated:

$$w_i^{RL} = \frac{\sum_{k=1}^{M_i^B} w_{ik}'^{RL}}{\sum_{k=1}^{M_i^B} \theta_{ik}^B} . \tag{9.6}$$

Step 4: Finally, we set $w_{ik}^{RL} = w_i^{RL}$ for all $k \in U_i^B$.

It should be noted that because they are present in the numerator and the denominator, the linkage weights do not need to be between 0 and 1. They just need to represent the likelihood of having a link between two units from populations U^A and U^B. It is also interesting to see that with $\tilde{\theta}_{j,ik} = \theta_{j,ik} / \theta_i^B$, where $\theta_i^B = \sum_{j=1}^{M^A} \sum_{k=1}^{M_i^B} \theta_{j,ik}$, we obtain an equivalent formulation to the one coming from the generalisation of the estimation weight described in section 4.5.

With the classical approach, each cluster *i* of U^B is assumed to have at least one non-zero link with a unit *j* of U^A. This constraint is translated here into the need of having, for each cluster *i* of U^B, at least one linkage weight $\theta_{j,ik}$ greater than zero with a unit *j* of U^A. In theory, it is not guaranteed that this constraint will be satisfied following the record linkage process. For example, it is possible that for a cluster *i* of U^B, there is no linkage weight $\theta_{j,ik}$ greater than zero. In that case, the estimation weight (9.6) underestimates the total Y^B. To solve this problem, the same solutions proposed in the context of the indicator variables $l_{j,ik}$ can be used. That is, two clusters can be collapsed, for example, in order to get at least one linkage weight $\theta_{j,ik}$ greater than zero for the new cluster. Unfortunately, this solution may require manual intervention, which has been avoided up to now by not using a decision rule. A better solution is to impute a link by choosing one link at random within the cluster. Then, a small value $\theta_{j,ik} > 0$ can be assigned arbitrarily for the chosen link.

Following the same steps as those from the proof of Theorem 4.1, the estimator $\hat{Y}^{RL,B}$ given by (9.4) can be rewritten in the following way:

$$\hat{Y}^{RL,B} = \sum_{j=1}^{M^A} \frac{t_j}{\pi_j^A} \sum_{i=1}^{N} \sum_{k=1}^{M_i^B} \theta_{j,ik} z_{ik}^{RL} = \sum_{j=1}^{M^A} \frac{t_j}{\pi_j^A} Z_j^{RL} \tag{9.7}$$

where $z_{ik}^{RL} = Y_i / \theta_i^B$ for all $k \in U_i^B$, and $\theta_i^B = \sum_{k=1}^{M_i^B} \sum_{j=1}^{M^A} \theta_{j,ik}$.

With this last expression, it can be shown that the estimator $\hat{Y}^{RL,B}$ is unbiased using the same development as Corollary 4.1. Finally, by following Corollary 4.2, the variance of $\hat{Y}^{RL,B}$ is given by

$$Var(\hat{Y}^{RL,B}) = \sum_{j=1}^{M^A} \sum_{j'=1}^{M^A} \frac{(\pi_{jj'}^A - \pi_j^A \pi_{j'}^A)}{\pi_j^A \pi_{j'}^A} Z_j^{RL} Z_{j'}^{RL}. \tag{9.8}$$

To estimate the variance (9.8), one of the two estimators (4.12) or (4.13) can be used by replacing the variable Z_j with Z_j^{RL}.

9.2.2 Approach 2: use all non-zero links above a given threshold

The use of the GWSM for all non-zero links might require the manipulation of very large files of size $M^A \times M^B$. This can occur if almost all pairs (j,ik) between populations U^A and U^B have non-zero linkage weights $\theta_{j,ik}$. In practice, even if this happens, it is strongly possible that most of these linkage weights will be very small or negligible. Even if the linkage weights are not non-zero, the links coming from these small linkage weights are probably not true. Indeed, looking at equation (9.2), we note that if $\theta_{j,ik}$ is very small, the conditional probability that there is a link between *j* and *ik* is then much smaller than the conditional probability that there is no link. In that case, it might be useful to only consider the links with linkage weights above a given threshold θ_{High}.

As with approach 1, we no longer use the indicator variables $l_{j,ik}$ identifying the links, but instead, we use the linkage weights $\theta_{j,ik}$ obtained in the first steps of the record linkage process. However, with approach 2, we restrict ourselves to the linkage weights greater than or equal to a threshold θ_{High}. The linkage weights below the threshold θ_{High} are considered as zeros. We therefore define the following linkage weight:

$$\theta_{j,ik}^{RLT} = \begin{cases} \theta_{j,ik} & \text{if } \theta_{j,ik} \geq \theta_{High} \\ 0 & \text{otherwise.} \end{cases} \tag{9.9}$$

For each unit *j* selected in s^A, we identify the units *ik* of U^B that have $\theta_{j,ik}^{RLT} > 0$. Let $\Omega^{RLT,B} = \{i \in U^B \mid \exists j \in s^A \text{ and } \theta_{j,i}^{RLT} > 0\}$ with $\theta_{j,i}^{RLT} = \sum_{k=1}^{M_i^B} \theta_{j,ik}^{RLT}$ be the set of the n^{RLT} clusters identified by the units $j \in s^A$, where "RLT" stands for record linkage with threshold. Note that $n^{RLT} \leq n^{RL}$. On the other hand, we have $n^{RLT} = n$ if the record linkage between U^A and U^B is done using the decision rule (9.3) with $\theta_{High} = \theta_{Low}$.

To estimate the total Y^B of population U^B, we can use the estimator

$$\hat{Y}^{RLT,B} = \sum_{i=1}^{n^{RLT}} \sum_{k=1}^{M_i^B} w_{ik}^{RLT} y_{ik}, \tag{9.10}$$

where w_{ik}^{RLT} is the estimation weight obtained from the GWSM. This weight is obtained by directly replacing the linkage weight $\theta_{j,ik}$ with the linkage weight $\theta_{j,ik}^{RLT}$ given by (9.9) in the steps of the GWSM described in section 9.2.1.

It is again interesting to see that with $\tilde{\theta}_{j,ik}^{RLT} = \theta_{j,ik}^{RLT} / \theta_i^{RLT,B}$, where $\theta_i^{RLT,B} = \sum_{j=1}^{M^A} \sum_{k=1}^{M_i^B} \theta_{j,ik}^{RLT}$, a formulation is obtained that is equivalent to the one coming from the generalisation of the estimation weight described in section 4.5.

By definition, the number of zero linkage weights $\theta_{j,ik}^{RLT}$ will be greater than or equal to the number of zero linkage weights $\theta_{j,ik}$. The constraint that each cluster *i* of U^B must have at least one linkage weight $\theta_{j,ik}^{RLT}$ greater than zero with a unit *j* of U^A will thus be more difficult to satisfy. To solve this problem, the same solutions proposed in section 9.2.1 can be used. For example, two clusters can be collapsed in the same way to get at least one linkage weight $\theta_{j,ik}^{RLT}$ greater than zero for each cluster *i* of $\Omega^{RLT,B}$. Unfortunately, this solution can require manual intervention, which has been avoided up to now by not using any decision rule. A better solution is to impute a link by randomly choosing a link within the cluster. A value of $\theta_{j,ik}^{RLT}$ equal to the threshold θ_{High} can then be assigned to this link.

As in section 9.2.1, the estimator $\hat{Y}^{RLT,B}$ given by (9.10) can be rewritten in the following way:

$$\hat{Y}^{RLT,B} = \sum_{j=1}^{M^A} \frac{t_j}{\pi_j^A} \sum_{i=1}^{N} \sum_{k=1}^{M_i^B} \theta_{j,ik}^{RLT} z_{ik}^{RLT} = \sum_{j=1}^{M^A} \frac{t_j}{\pi_j^A} Z_j^{RLT}, \tag{9.11}$$

where $z_{ik}^{RLT} = Y_i / \theta_i^{RLT,B}$ for all $k \in U_i^B$, and $\theta_i^{RLT,B} = \sum_{k=1}^{M_i^B} \sum_{j=1}^{M^A} \theta_{j,ik}^{RLT}$.

With (9.11), the estimator $\hat{Y}^{RLT,B}$ can be proven to be unbiased using the same development as Corollary 4.1. Finally, by following Corollary 4.2, the variance of $\hat{Y}^{RLT,B}$ is given by

$$Var(\hat{Y}^{RLT,B}) = \sum_{j=1}^{M^A}\sum_{j'=1}^{M^A} \frac{(\pi_{jj'}^A - \pi_j^A \pi_{j'}^A)}{\pi_j^A \pi_{j'}^A} Z_j^{RLT} Z_{j'}^{RLT} . \quad (9.12)$$

To estimate the variance (9.12), one of the two estimators (4.12) or (4.13) can be used by replacing the variable Z_j with Z_j^{RLT}.

9.2.3 Approach 3: choose the links randomly

In order to avoid making a decision on the links between units j from U^A and units k of clusters i from U^B, one can decide to simply choose the links at random from the set of all links with linkage weights $\theta_{j,ik}$ greater than zero. For this, it is reasonable to choose the links with probabilities proportional to the linkage weights. This can be done using *Bernoulli trials* where, for each pair(j,ik), we can decide to accept a link or not by generating a random number $\nu_{j,ik} \sim \text{U}(0,1)$ that is then compared to a quantity proportional to the linkage weight $\theta_{j,ik}$.

In the point of view of record linkage, this approach cannot be considered as optimal. Indeed, when using the decision rule (9.3) of Fellegi and Sunter (1969), the idea is to minimise the number of false links and false non-links. The link $l_{j,ik}$ is accepted only if the linkage weight $\theta_{j,ik}$ is large (i.e., $\theta_{j,ik} \geq \theta_{High}$), or if it is moderately large (i.e., $\theta_{Low} < \theta_{j,ik} < \theta_{High}$) and has been accepted after manual resolution. The random selection of links using Bernoulli trials can lead to the selection of links that would have not been accepted through the decision rule (9.3), even though the selection probabilities are proportional to the linkage weights. Following the Bernoulli trials, some of the links accepted between the two populations U^A and U^B can be false, and some other links may have been falsely rejected. The linkage errors therefore tend to be higher if the Bernoulli trials are used. However, in the present context, the quality of the links can be considered as a secondary interest. The problem here is to estimate the total Y^B using the sample s^A selected from U^A, and not to evaluate the quality of the links. In section 9.3, the precision of the estimates of Y^B will be measured with respect to the sampling variability of the estimators, by conditioning on the linkage weights $\theta_{j,ik}$. Note that this sampling variability will take into account the random selection of the links, but not the linkage errors.

To reduce the number of non-zero links, the present approach is therefore considered as being of potential interest, even if the quality of the resulting links can be questionable.

The first step before performing the Bernoulli trials is to transform the linkage weights in a way such that they are contained in the [0,1] interval. By looking at the definition (9.1), it can be seen that the linkage weights $\dot{\theta}_{j,ik}$ correspond in fact to a "logit" transformation (in base 2) of the probability $P(\upsilon_{j,ik} \mid \Delta_{1j,ik}\Delta_{2j,ik}...\Delta_{pj,ik})$. In the same way, the linkage weights $\theta_{j,ik}$ given by (9.2) depend only on this same probability. Hence, one way to transform the linkage weights is simply to use the probability $P(\upsilon_{j,ik} \mid \Delta_{1j,ik}\Delta_{2j,ik}...\Delta_{pj,ik})$. From (9.1), we obtain this result by using the function $\tilde{\tilde{\theta}} = 2^{\dot{\theta}}/(1+2^{\dot{\theta}})$ and, from (9.2), by using $\tilde{\tilde{\theta}} = \theta/(1+\theta)$. If the linkage weight are not obtained through a definition similar to (9.1) or (9.2), another possible transformation is to simply divide each weight by the maximum value θ_{Max} = $\max(\theta_{j,ik} \mid j=1,...M^A, i=1,...,N, k=1,...,M_i^B)$. Note that we assume here that the linkage weights are all greater than or equal to zero, which is the case from definition (9.2), but not necessarily in general.

Once the adjusted linkage weights $\tilde{\tilde{\theta}}_{j,ik}$ have been obtained, we generate for each pair (*j,ik*) a random number $\nu_{j,ik} \sim \mathrm{U}(0,1)$. Then, we assign the value 1 to the indicator variable $\tilde{\tilde{l}}_{j,ik}$ if $\nu_{j,ik} \le \tilde{\tilde{\theta}}_{j,ik}$, and the value 0 otherwise. This process provides a set of links similar to the ones used in the classical approach, with the exception that now the links are determined randomly and not through a decision process like (9.3). Note that since $E(\tilde{\tilde{l}}_{j,ik}) = \tilde{\tilde{\theta}}_{j,ik}$, the sum of the adjusted linkage weights $\tilde{\tilde{\theta}}_{j,ik}$ corresponds to the expected total number of links L from the Bernoulli trials, i.e.,

$$\sum_{j=1}^{M^A}\sum_{i=1}^{N}\sum_{k=1}^{M_i^B}\tilde{\tilde{\theta}}_{j,ik} = L. \tag{9.13}$$

With the present approach, by randomly selecting links, it is strongly possible that Constraint 2.1 related to the GWSM will not be satisfied. To correct this problem, a link can be imputed by choosing the link with the largest linkage weight $\tilde{\tilde{\theta}}_{j,ik}$ within the cluster. The link can also be selected randomly with a probability proportional to $\tilde{\theta}_{j,ik}$.

For each unit *j* selected in s^A, we identify here the units *ik* of U^B that have $\tilde{\tilde{l}}_{j,ik} = 1$. Let $\tilde{\tilde{\Omega}}^B = \{i \in U^B \mid \exists j \in s^A \text{ and } \tilde{\tilde{L}}_{j,i} > 0\}$ where $\tilde{\tilde{L}}_{j,i} = \sum_{k=1}^{M_i^B} \tilde{\tilde{l}}_{j,ik}$ be the set of the $\tilde{\tilde{n}}$ clusters identified by the units $j \in s^A$. Note that $\tilde{\tilde{n}} \le n^{RL}$. Unfortunately, in contrast to n^{RL} and n^{RLT}, the number of clusters $\tilde{\tilde{n}}$ is hardly comparable to *n*, the number of clusters obtained using the classical approach.

To estimate the total Y^B of the population U^B, we can use

$$\tilde{\tilde{Y}}^B = \sum_{i=1}^{\tilde{\tilde{n}}} \sum_{k=1}^{M_i^B} \tilde{\tilde{w}}_{ik} y_{ik} \tag{9.14}$$

where $\tilde{\tilde{w}}_{ik}$ is the estimation weight obtained from the GWSM. This weight is obtained by directly replacing the indicator variables $l_{j,ik}$ with $\tilde{\tilde{l}}_{j,ik}$ in the steps of the GWSM described in section 2.1.

Steps of the GWSM for approach 3

Step 1: For each unit *k* of the clusters *i* from $\tilde{\tilde{\Omega}}^B$, the initial weight $\tilde{\tilde{w}}'_{ik}$ is calculated, that is:

$$\tilde{\tilde{w}}'_{ik} = \sum_{j=1}^{M^A} \tilde{\tilde{l}}_{j,ik} \frac{t_j}{\pi_j^A}, \tag{9.15}$$

where $t_j = 1$ if $j \in s^A$, and 0 otherwise.

Step 2: For each unit *k* of the clusters *i* from $\tilde{\tilde{\Omega}}^B$, $\tilde{\tilde{L}}_{ik}^B = \sum_{j=1}^{M^A} \tilde{\tilde{l}}_{j,ik}$ is calculated. The quantity $\tilde{\tilde{L}}_{ik}^B$ represents the realised number of links between the units of U^A and unit *k* of cluster *i* from U^B.

Step 3: The final weight $\tilde{\tilde{w}}_i$ is calculated:

$$\tilde{\tilde{w}}_i = \frac{\sum_{k=1}^{M_i^B} \tilde{\tilde{w}}'_{ik}}{\sum_{k=1}^{M_i^B} \tilde{\tilde{L}}_{ik}^B}. \tag{9.16}$$

Step 4: Finally, we set $\tilde{\tilde{w}}_{ik} = \tilde{\tilde{w}}_i$ for all $k \in U_i^B$.

By conditioning on the accepted links $\tilde{\tilde{l}}_{j,ik}$, it can be shown that the estimator $\hat{\tilde{Y}}^B$ given by (9.14) is unbiased, assuming of course that Constraint 2.1 is satisfied. Let $E_l(.)$ be the expected value carried out in relation to all the possible realisations of links. Let $\tilde{\tilde{\mathbf{L}}}$ be the set of realised links, i.e.,

$$\tilde{\tilde{\mathbf{L}}} = \left\{ \tilde{\tilde{l}}_{j,ik} \right\}_{j=1,i=1,k=1}^{M^A,N,M_i^B}.$$

We then have

$$E(\hat{\tilde{Y}}^B) = E_l[E(\hat{\tilde{Y}}^B \mid \tilde{\tilde{\mathbf{L}}})]. \tag{9.17}$$

By conditioning on the set $\tilde{\tilde{\mathbf{L}}}$, the estimator (9.14) is then equivalent to the estimator (2.1) (or the estimator (4.1)). From Corollary 4.1, $E(\hat{\tilde{Y}}^B \mid \tilde{\tilde{\mathbf{L}}}) = Y^B$ is directly obtained and therefore, the estimator (9.14) is conditionally unbiased. Consequently, this estimator is unbiased in an unconditional way. To obtain the variance of $\hat{\tilde{Y}}^B$, we again proceed in a conditional way from

$$Var(\hat{\tilde{Y}}^B) = E_l[Var(\hat{\tilde{Y}}^B \mid \tilde{\tilde{\mathbf{L}}})] + Var_l[E(\hat{\tilde{Y}}^B \mid \tilde{\tilde{\mathbf{L}}})].$$

First of all, since $E(\hat{\tilde{Y}}^B \mid \tilde{\tilde{\mathbf{L}}}) = Y^B$, we have

$$Var_l[E(\hat{\tilde{Y}}^B \mid \tilde{\tilde{\mathbf{L}}})] = Var_l[Y^B] = 0. \tag{9.18}$$

By conditioning on the set $\tilde{\tilde{\mathbf{L}}}$, it was already mentioned that the estimator (9.14) is equivalent to the estimator (2.1). By Corollary 4.2, the following result is thus obtained:

$$Var(\hat{\tilde{Y}}^B \mid \tilde{\tilde{\mathbf{L}}}) = \sum_{j=1}^{M^A}\sum_{j'=1}^{M^A} \frac{(\pi_{jj'}^A - \pi_j^A \pi_{j'}^A)}{\pi_j^A \pi_{j'}^A} \tilde{\tilde{Z}}_j \tilde{\tilde{Z}}_{j'} \tag{9.19}$$

where $\tilde{\tilde{Z}}_j = \sum_{i=1}^{N} \sum_{k=1}^{M_i^B} \tilde{\tilde{l}}_{j,ik} \tilde{\tilde{z}}_{ik}$ with $\tilde{\tilde{z}}_{ik} = Y_i / \tilde{\tilde{L}}_i^B$. The variance of $\hat{\tilde{Y}}^B$ can therefore be written

$$Var(\hat{\tilde{Y}}^B) = E_l\left[\sum_{j=1}^{M^A}\sum_{j'=1}^{M^A} \frac{(\pi_{jj'}^A - \pi_j^A \pi_{j'}^A)}{\pi_j^A \pi_{j'}^A} \tilde{\tilde{Z}}_j \tilde{\tilde{Z}}_{j'} \right]. \tag{9.20}$$

To estimate the variance (9.20), one of the two estimators (4.12) or (4.13) can be used by replacing the variable Z_j with $\tilde{\tilde{Z}}_j$.

9.2.4 Some remarks

The three proposed approaches do not use the decision rule (9.3). They also do not require any manual resolution. Consequently, the answer to question (c) is yes. That is, the GWSM can help in reducing the manual resolution required by record linkage. Note that there is however a price to pay for avoiding manual resolution.

First, with approach 1, the number n^{RL} of clusters identified by the units $j \in s^A$ is greater than or equal to the number n of clusters identified by the classical approach, i.e., when the decision rule (9.3) is used to accept the links or not. This happens because we use all non-zero links, and not just the ones satisfying the decision rule (9.3). As a consequence, the collection costs with approach 1 are greater than or equal to the ones related to the classical approach. It needs then to be checked which ones are the most important: the collection costs or the costs of manual resolution. Note that if the precision resulting from the use of approach 1 is much higher than the one from the classical approach, it can be more advantageous to choose approach 1 than the classical approach.

With approach 2, we have $n^{RLT} \leq n^{RL}$ and therefore the collection costs of this approach are less than or equal to the ones of approach 1. If the precision of approach 2 is comparable to the one of approach 1, then approach 2 will certainly be more advantageous than approach 1. By comparing approach 2 with the classical approach, it can be seen that the collection costs can be almost equivalent if the value of the threshold θ_{High} is chosen to be relatively close to the lower and upper thresholds of the decision rule (9.3).

Note that approach 2 does not use any manual resolution. If the precision of approach 2 is at least comparable to the one of the classical approach, then approach 2 is more advantageous. Note that if $\theta_{High} = \theta_{Low}$, the two approaches differ only in the definition of the estimation weights ensuing from the GWSM. Approach 2 uses the linkage weights $\theta_{j,ik}^{RLT}$, while the classical approach uses the indicator variables $l_{j,ik}$. Setting $\theta_{High} = \theta_{Low}$, it is certainly of interest to know which approach has the highest precision.

With approach 3, the number of selected links is less than or equal to the number of non-zero links used by approach 1, i.e., $\tilde{\tilde{n}} \leq n^{RL}$. Hence, the collection costs of approach 3 are less than or equal to the ones of approach 1. As mentioned before, unlike n^{RL} and n^{RLT}, the number of clusters $\tilde{\tilde{n}}$ is hardly comparable to *n*. The two depend on different parameters: the classical approach depends on the thresholds θ_{Low} and θ_{High}, while approach 3 depends on the adjusted linkage weights $\tilde{\tilde{\theta}}_{j,ik}$.

9.3 SIMULATION STUDY

We performed a simulation study to evaluate the approaches presented in this chapter, including the classical approach. For this study, we compared the precision obtained for the estimation of the total Y^B for five different approaches:

Approach 1: use all non-zero links with the linkage weights $\theta_{j,ik}$

Approach 2: use all non-zero links above a threshold

Approach 3: choose the links randomly

Approach 4: classical approach

Approach 5: use all non-zero links with the indicator variables $l_{j,ik}$.

Approach 5 is a mixture of approach 1 and the classical approach. It consists of first accepting all links of the pairs *(j,ik)* that have a linkage weight greater than zero, i.e., assign $l_{j,ik} = 1$ for all pairs *(j,ik)* where $\theta_{j,ik} > 0$, and $l_{j,ik} = 0$ otherwise. The GWSM described in chapter 2 is then used to estimate the total Y^B from the estimator (2.1). Approach 5 was added to the simulations to verify the effect of using the indicator variables $l_{j,ik}$ instead of the linkage weights $\theta_{j,ik}$ when using all non-zero links. As with all the other approaches, according to Corollary 4.1, approach 5 is unbiased. Since the five approaches yield unbiased estimates of the total Y^B, we compared them with the *standard error* (the square root of the variance), and more specifically with the *coefficient of variation* (the standard error divided by the expected value of the estimator).

9.3.1 Data used

The simulation study was performed using the agricultural data presented in section 7.4.4. Thus, the study again is inspired by Statistics Canada's Whole Farm Data Base. Recall that this database has information on livestock, crops and the income and expenditures (tax data) of Canadian farms (Statistics Canada, 2000a). The data used for the simulations come from Québec and New Brunswick. Although the simulations were inspired by the Whole Farm Data Base, some processes and data were changed for reasons of confidentiality, and also to not needlessly complicate the discussion. However, we believe that these changes do not affect the conclusions drawn from the simulations.

The population U^A is a list of M^A farms coming from the 1996 Farm Register. This list essentially comes from the 1991 Canadian Census of Agriculture, with different updates that have been made since 1991. The units j from U^A thus represent farms, but note that each farm j can have many farm operators. In addition to the FarmID, the Farm Register contains a farm operator number together with some demographic variables related to the farm operators.

The target population U^B is a list of M^B tax records (or income tax reports) from the Canadian Revenue Agency (CRA). This second list is the 1996 CRA Unincorporated Business File that contains tax data for the persons declaring at least one farming income. This file contains a household number (only for a sample), a tax filer number, and also demographic variables related to the tax filers. The units k are thus the tax reports that are completed by the different members of households i (or clusters). The target population U^B has N households. The respective sizes of the populations U^A and U^B are given in Table 7.1.

For the simulations, linkage has been performed for the two populations U^A and U^B (in fact, linkage of the files A and B related to these populations). To do this, a linkage process was used based on the matching of five variables. It was performed using the MERGE statement in SAS®. The records on both files were compared to one another in order to determine whether or not there is a match. The record linkage was performed using the following five key variables common to both files:

1) first name (modified using NYSIIS)
2) last name (modified using NYSIIS)
3) birth date

4) street address
5) postal code.

The first name and last name were modified using the *NYSIIS* system. This basically changes the name in phonetic expressions, which in turn increases the chance of finding matches by reducing the probability that a good match is rejected because of a spelling mistake or a typing error.

Records that matched on all five variables were given the highest linkage weight ($\theta = 60$). Records that matched on only a subset of at least two of the five variables received a lower non-zero linkage weight ($\theta = 2$). Pairs of records that did not match on any combination of key variables were considered as pairs having no possible links, which is equivalent to having a linkage weight of zero.

Two different threshold were chosen for the simulations: $\theta_{High} = \theta_{Low} = 15$ and $\theta_{High} = \theta_{Low} = 30$. The upper and lower thresholds, θ_{High} and θ_{Low}, were set to be the same to avoid the grey area where some manual intervention is needed when applying the decision rule (9.3).

Following the linkage process, the constraint requiring that each cluster i of the target population U^B have at least one non-zero link was not satisfied for all clusters. To correct the situation, we imputed a link by choosing the link with the largest linkage weight $\theta_{j,ik}$ within the cluster. In the case where all linkage weights are zero, we chose a link at random.

The record linkage process used here does not exactly correspond to the one used to construct the Whole Farm Data Base. For more information on the exact process, refer to Lim (2000). We believe that the changes, however, do not affect the conclusions drawn from the simulations. Recall that the main goal of the simulations is to evaluate the different approaches for the estimation of Y^B, and not to solve the problems related to the construction of the Whole Farm Data Base.

Following record linkage, it turns out that the populations U^A and U^B are linked by complex links. Indeed, a farm j sometimes has many operators and each operator returns one tax report k to the CRA. There is then a "one-to-many" link since we have one farm j linked to many tax reports k. On the other hand, an operator who deals with more than one farm j can return a single tax report k for the set of farms that he operates. Therefore, this type of link is "many-to-one" since there are many farms j

linked to a single tax report k. Finally, there are also situations of complex links where the operators deal with more than one farm and where each farm has a number of different operators. The populations U^A and U^B as well as their links can be represented by Figure 2.1.

9.3.2 Sampling plan

For the simulations, the sample s^A was selected from U^A (Farm Register) using simple random sampling without replacement, without any stratification. We also considered two sampling fractions: 30% and 70%. The variable of interest y for which we want to estimate the total Y^B is the total farming income. Since we have the entire populations of farms and tax records, it was possible to calculate the value of Y^B and the variances from the theoretical formulas developed for this approach. Furthermore, because a simple random sampling without replacement was performed, these theoretical formulas can be simplified. For example, in the case of approach 1, the variance of $\hat{Y}^{RL,B}$ given by (9.8) can then be written in the following form:

$$Var(\hat{Y}^{RL,B}) = M^A \frac{(1-f^A)}{f^A} \frac{1}{M^A-1} \sum_{j=1}^{M^A} (Z_j^{RL} - \overline{Z}^{RL})^2 , \qquad (9.21)$$

where $f^A = m^A / M^A$ is the *sampling fraction* and $\overline{Z}^{RL} = \frac{1}{M^A} \sum_{j=1}^{M^A} Z_j^{RL}$.

A *Monte Carlo study* was also conducted to empirically calculate the bias and the variance under the different approaches. Note that for approach 3, only the Monte Carlo study was used. For the Monte Carlo study, 500 samples s^A from U^A were selected for each sampling fraction 30% and 70%, and for each threshold 15 and 30. The empirical bias and the empirical variance of each estimator (represented by $\hat{Y}$) were calculated using

$$\hat{Bias}(\hat{Y}) = \hat{E}(\hat{Y}) - Y^B = \frac{1}{500} \sum_{s^A=1}^{500} \hat{Y}_{s^A} - Y^B \qquad (9.22)$$

$$\hat{Var}(\hat{Y}) = \frac{1}{500} \sum_{s^A=1}^{500} (\hat{Y}_{s^A} - \hat{E}(\hat{Y}))^2 . \qquad (9.23)$$

The coefficients of variation (CV) were then calculated by using

$$\hat{CV}(\hat{Y}) = 100 \times \frac{\sqrt{\hat{Var}(\hat{Y})}}{\hat{E}(\hat{Y})}. \tag{9.24}$$

The Monte Carlo study was, among other things, performed to verify in an empirical manner the accuracy of the theoretical formulas given in section 9.2. The results all indicated that the theoretical formulas are exact.

9.3.3 Results and discussion

The results of the simulations are given in Figures 9.1 to 9.4, Table 9.1 and Figure 9.5. Figures 9.1 to 9.4 provide histograms of the CVs obtained for each of the five approaches. Eight histograms are shown, corresponding to the eight cases obtained by crossing the two provinces Québec and New Brunswick, the two sampling fractions 30% and 70%, and the two thresholds 15 and 30.

On each bar of the histograms, one can see the number of non-zero links between U^A and U^B for each of the five approaches. For approach 3, it is in fact the expected number of non-zero links. Note that the number (expected or not) of non-zero links does not change from one sampling fraction to another.

Table 9.1 shows, for each of the eight cases, the average number of clusters surveyed for each approach. This average is calculated with respect to the 500 samples s^A used for the simulations. The numbers is parentheses represent the standard error for the number of surveyed clusters. The standard errors are relatively small compared to the averages and therefore, the number of clusters surveyed do not vary greatly from one sample to another.

Figure 9.5 gives, for each of the eight cases, a graph of the obtained CVs for the five approaches as a function of the average number of surveyed clusters.

By looking at Figures 9.1 to 9.4, it can be seen that in all cases, approaches 1 and 5 give the smallest CVs for the estimation of total farming income. Therefore, using all non-zero links produces estimates with the greatest precision. Looking at Table 9.1, we note however that these approaches are the ones for which the number of surveyed clusters is the highest. In fact, we can see that the greater the number of surveyed clusters, the greater the precision of the estimates is. This result is shown in Figure 9.5 where we can see that the CVs tend to decrease as the average number of surveyed clusters increases. Although this observation is well known in the classical sampling theory, it is not necessarily evident in the context of indirect sampling. As we can see from equations (4.11a) and (4.11b), it is not the sample size of s^A that increases, but rather the homogeneity of the derived variables Z_j.

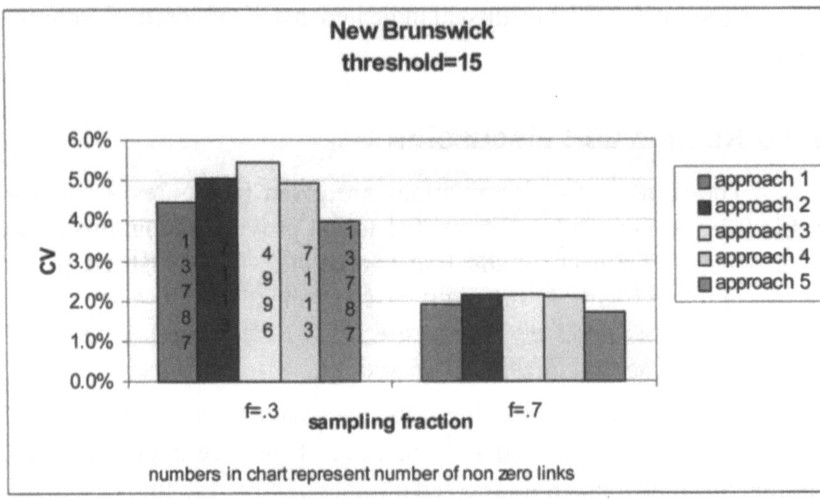

Figure 9.1: *CVs for New Brunswick (with* $\theta_{High} = \theta_{Low}$ *=15)*

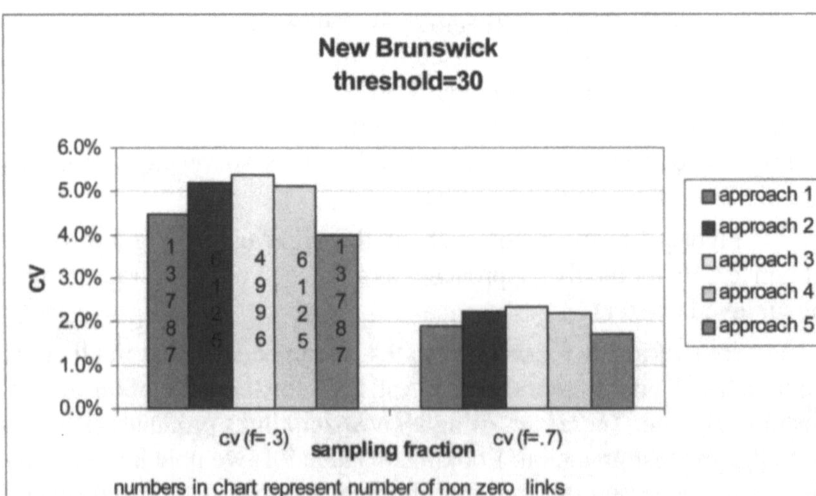

Figure 9.2: *CVs for New Brunswick (with* $\theta_{High} = \theta_{Low}$ =30)

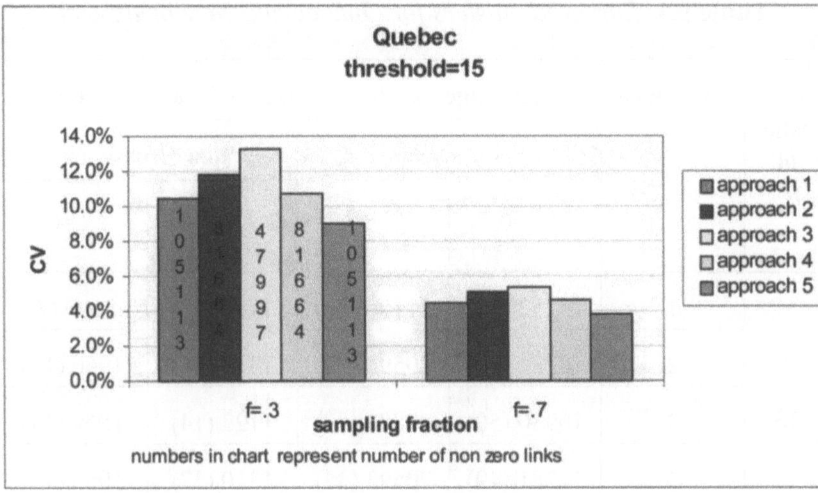

Figure 9.3: *CVs for Québec (with* $\theta_{High} = \theta_{Low}$ *=15)*

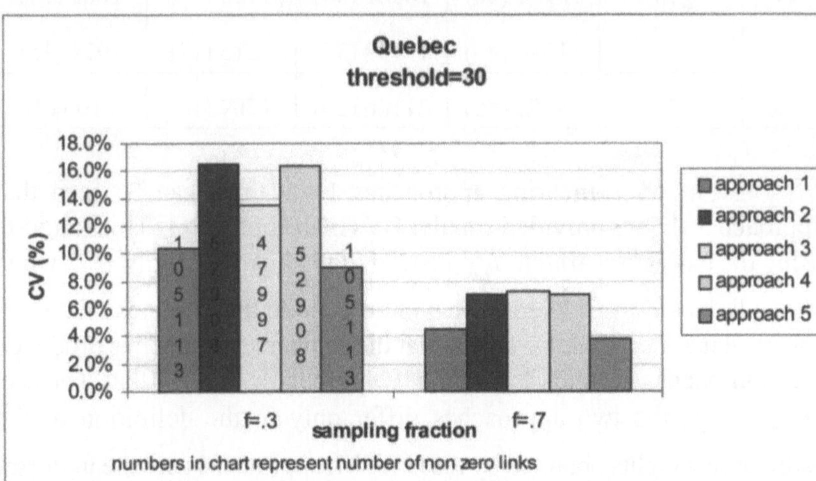

Figure 9.4: *CVs for Québec (with* $\theta_{High} = \theta_{Low}$ =30)

Table 9.1: *Surveyed clusters for Québec and New Brunswick*

Threshold θ_{High}	Approach	Average number of surveyed clusters (s.e.)			
		Québec		New Brunswick	
		f^A=0.3	f^A=0.7	f^A=0.3	f^A=0.7
15	1	15752 (58)	21106 (30)	1709 (18)	2100 (7)
	2	14281 (49)	20593 (34)	1310 (17)	1966 (13)
	3	10930 (50)	18881 (47)	1123 (14)	1869 (14)
	4	14281 (49)	20593 (34)	1310 (17)	1966 (13)
	5	15752 (58)	21106 (30)	1709 (18)	2100 (7)
30	1	15752 (58)	21106 (30)	1709 (18)	2100 (7)
	2	11310 (45)	19139 (37)	1215 (17)	1924 (15)
	3	10930 (50)	18881 (47)	1123 (14)	1869 (14)
	4	11310 (45)	19139 (37)	1215 (17)	1924 (15)
	5	15752 (58)	21106 (30)	1709 (18)	2100 (7)

Now, by comparing approaches 1 and 5, it can be seen that approach 5 always provided smaller CVs than approach 1. This suggests using the indicator variable $l_{j,ik}$ instead of the linkage weight $\theta_{j,ik}$ when all the links are considered to be non-zero. Note that it seems this result can be generalised when we note that the same phenomenon is produced for approaches 2 and 4 (classical approach). Recall that because $\theta_{High} = \theta_{Low}$, the two approaches differ only in the definition of the estimation weights obtained by the GWSM; approach 4 uses the indicator variable $l_{j,ik}$ and approach 2, the linkage weight $\theta_{j,ik}$. This result is particularly important because it corresponds to the conclusions of Kalton and Brick (1995) and the ones in section 4.6.3, namely that by using $\tilde{\theta}_{j,ik} = l_{j,ik} / L_i^B$ in the version of the GWSM described in section 4.5, we should then approach minimal variances for the estimation of the total Y^B.

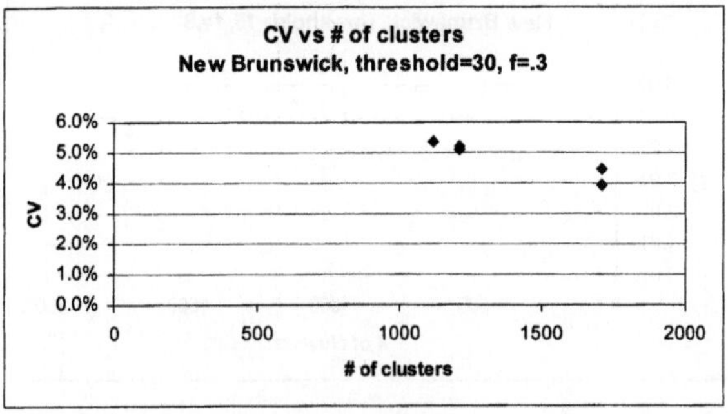

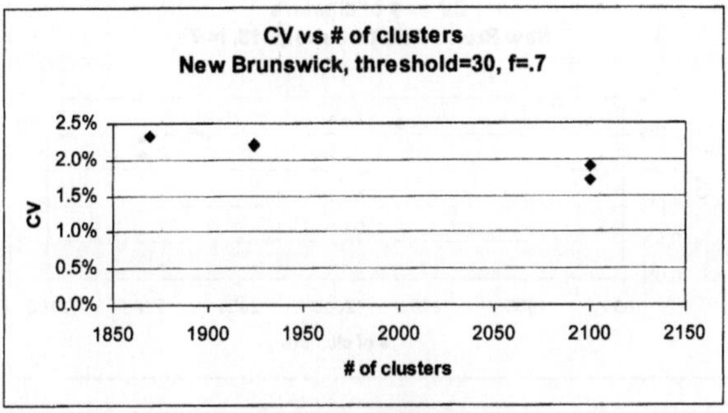

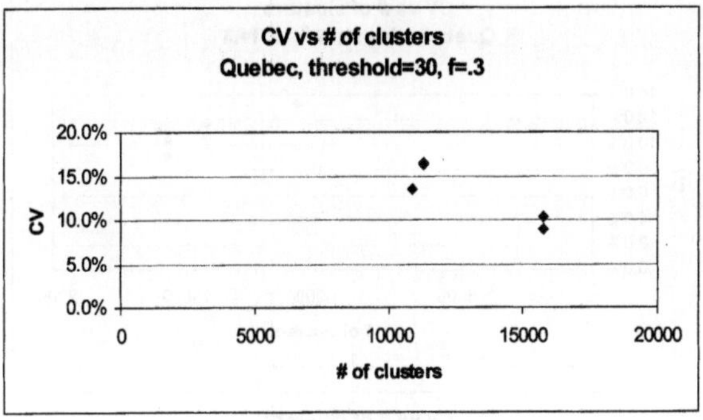

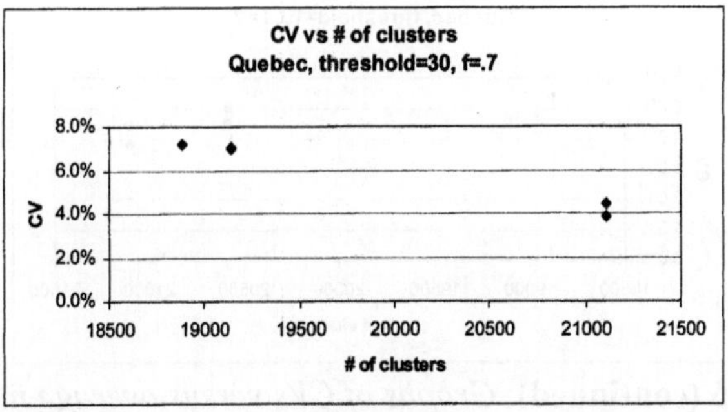

Figure 9.5: *Graphs of CVs versus average number of surveyed clusters*

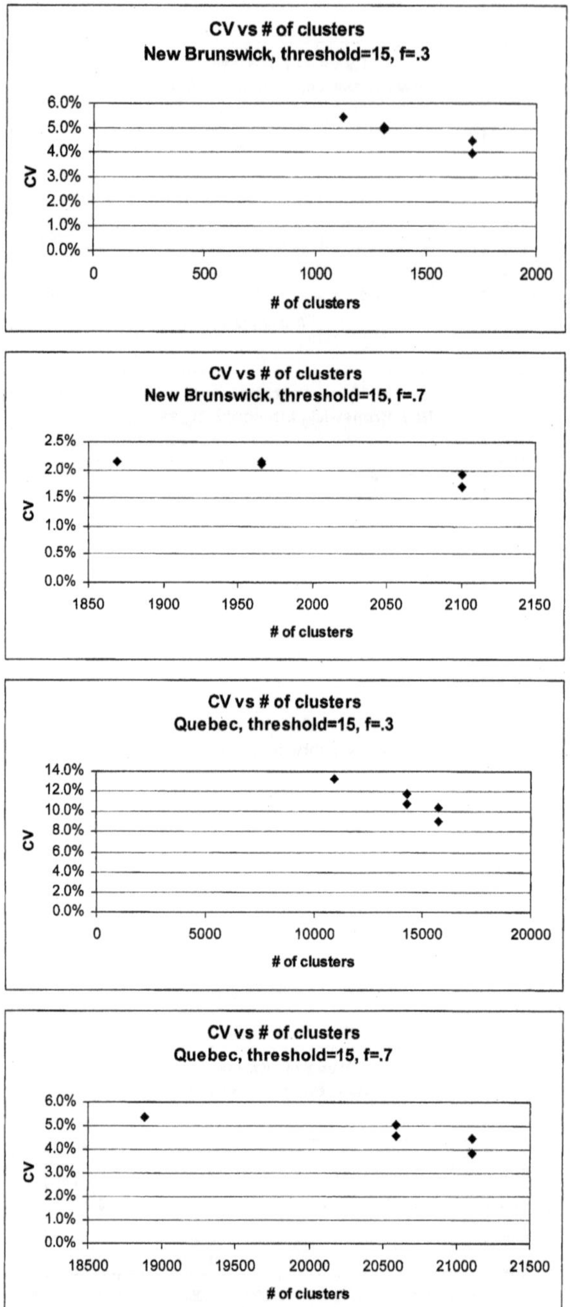

Figure 9.5 (continued): *Graphs of CVs versus average number of surveyed clusters*

Now consider approach 3. For seven out of the eight histograms from Figures 9.1 to 9.4, approach 3 produced the highest CVs. It should be noted however that this approach is the one that is based on the lowest number of non-zero links, and also the lowest number of surveyed clusters. Therefore, the poor performance of approach 3 is not surprising.

Recall that the number of non-zero links used by approach 3 does not depend on the threshold θ_{High}, and thus the CVs obtained for thresholds 15 and 30 are the same. For $\theta_{High}=15$, the CV obtained for Québec for approach 3 proves to be higher than the ones obtained for approaches 2 and 4, and these two approaches use more non-zero links and more surveyed clusters. For $\theta_{High}=30$, the CV obtained for approach 3 proves to be lower than the ones obtained for approaches 2 and 4, but these two approaches still used more non-zero links and more surveyed clusters. Therefore, there are intermediate situations where, with $15<\theta_{High}<30$, we get equal CVs for approaches 3 and 2, and equal CVs for approaches 3 and 4. As a result, to get equal CVs for approach 3 and each of approaches 2 and 4, more links (and more surveyed clusters) must be used by approaches 2 and 4. This suggests that approach 3 can, in some cases, be more worthwhile than approaches 2 and 4 because it produces estimates with the same precision but with lower collection costs.

So as to better compare approach 3 and approaches 2 and 4, we made the expected number of non-zero links to be the same as the number of non-zero links used by approaches 2 and 4. To do this, we have transformed the linkage weights $\theta_{j,ik}$ into new weights $\tilde{\tilde{\theta}}_{j,ik}$ such that

$$\sum_{j=1}^{M^A}\sum_{i=1}^{N}\sum_{k=1}^{M_i^B}\tilde{\tilde{\theta}}_{j,ik}=L_0\,, \tag{9.25}$$

where L_0 is the desired number of non-zero links. The transformation used was the following:

$$\tilde{\tilde{\theta}}_{j,ik}=\begin{cases}\theta_{j,ik}/\theta_{\bullet} & \text{if } \theta_{j,ik}/\theta_{\bullet}\le 1\\ 1 & \text{otherwise}\end{cases} \tag{9.26}$$

where $\theta_{\bullet}$ was determined iteratively so that constraint (9.25) is satisfied. The use of approach 3 with the transformed linkage weights by (9.26) was called approach 6. The results of the simulations are presented in Figures 9.6 to 9.9.

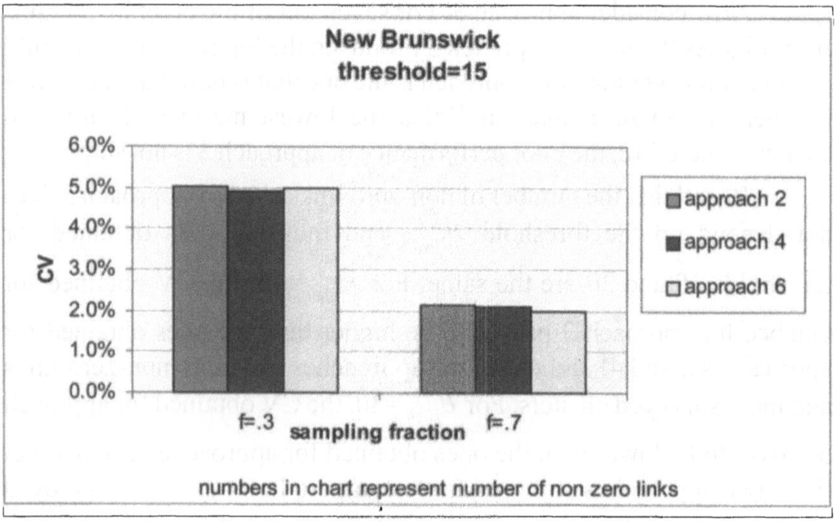

Figure 9.6: *CVs for New Brunswick* (*with* $\theta_{High} = \theta_{Low} = 15$)

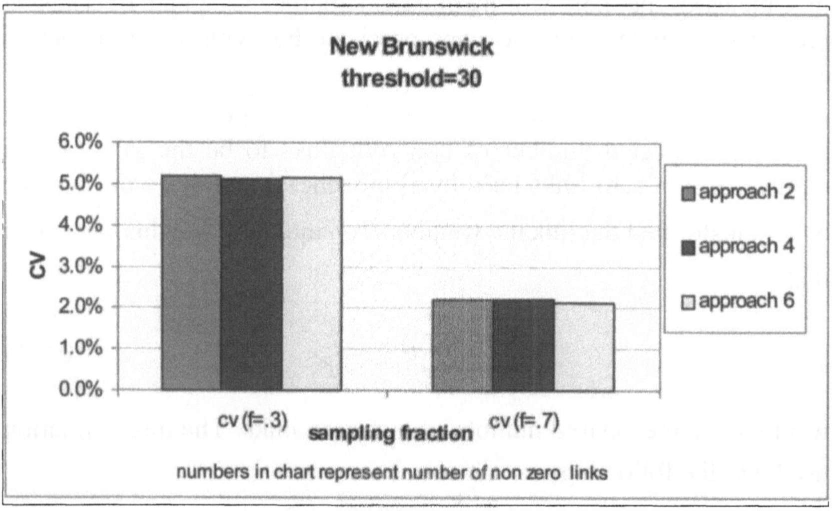

Figure 9.7: *CVs for New Brunswick* (*with* $\theta_{High} = \theta_{Low} = 30$)

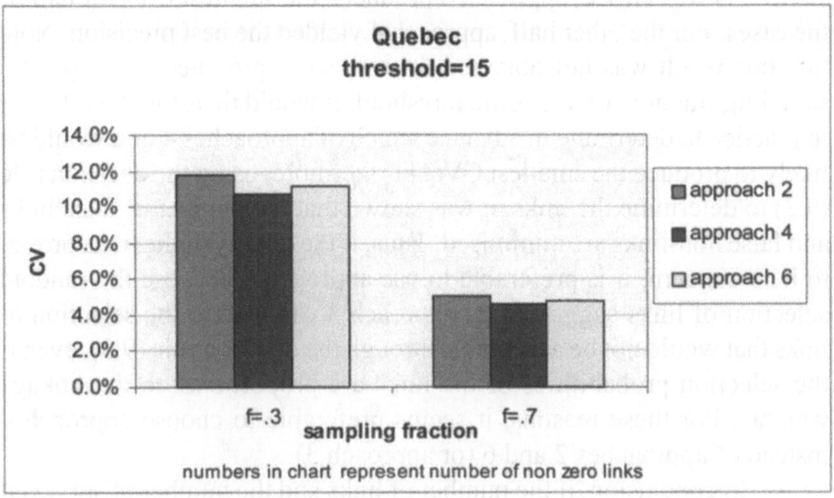

Figure 9.8: *CVs for Québec (with* $\theta_{High} = \theta_{Low} = 15$)

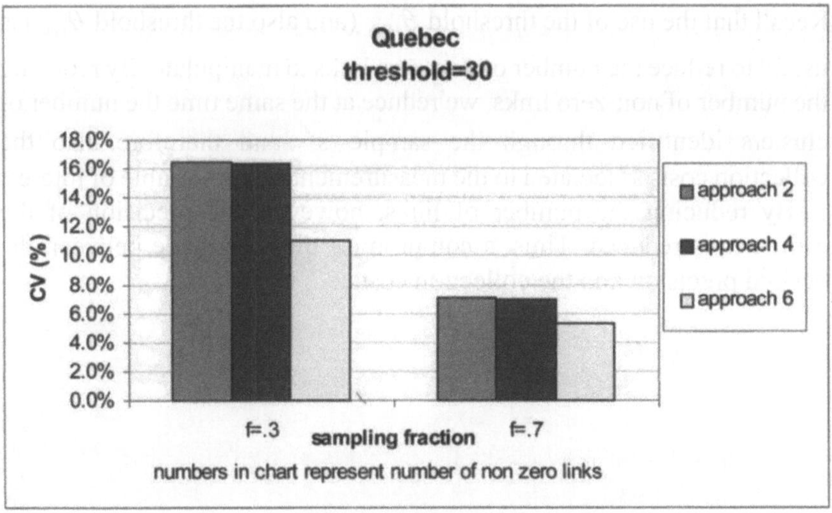

Figure 9.9: *CVs for Québec (with* $\theta_{High} = \theta_{Low} = 30$)

As we can see, approach 6 produced the smallest CVs for half of the cases. For the other half, approach 4 yielded the best precision. Note that this result was not obtained for a specific province, or a specific sampling fraction, or a specific threshold. It would therefore be difficult in practice to determine in advance which of approaches 4 or 6 would be likely to produce the smallest CVs. Furthermore, using the decision rule (9.3) to determine the links, it was shown that the number of false links and false non-links are minimised. Thus, if the quality of the links proves to be a concern, it is preferable to use approach 4 because the random selection of links suggested by approach 3 can lead to the selection of links that would not be acceptable through the decision rule (9.3), even if the selection probabilities of the links are proportional to the linkage weights. For these reasons, it seems preferable to choose approach 4 instead of approaches 2 and 6 (or approach 3).

In conclusion, if the number of links and the number of surveyed clusters do not pose a problem, it is suggested to use approach 5, i.e., to consider all the links of pairs (j,ik) that have a linkage weight $\theta_{j,ik}$ greater than zero, and to use the GWSM described in chapter 2 to estimate the total Y^B from the estimator (2.1). If the number of surveyed clusters proves to be too large because, for example, it leads to collection costs that are too high, approach 4 can be seen as a reasonable choice. Recall that the use of the threshold θ_{High} (and also the threshold θ_{Low}) is useful to reduce the number of non-zero links to manipulate. By reducing the number of non-zero links, we reduce at the same time the number of clusters identified through the sample s^A and therefore also the collection costs associated to the measurement of the variable of interest y. By reducing the number of links, however, the precision of the estimates is reduced. Thus, a compromise must be made between the desired precision and the collection costs.

CHAPTER 10

CONCLUSION

Throughout this book, it was shown that indirect sampling is a convenient way to obtain a sample to produce estimates for a target population U^B, when the only frame available is one for a population U^A related to U^B. Once the sample is selected, the GWSM can prove to be a viable solution for producing estimation weights in the context of indirect sampling. These weights lead to unbiased estimates. They roughly correspond to an average of the sampling weights for the units of the population U^A from which the sample is selected. Recall that the GWSM works even if the links between the two populations U^A and U^B are complex, that is, they are of the type "many-to-many."

The GWSM turns out to be particularly useful because it provides:

1) a weighting for an indirect sampling meant for rare populations;
2) a weighting using only the selection probabilities of selected units;
3) a weighting for populations related by complex links;
4) a weighting for non-linked units.

It was mentioned in the introduction that the GWSM was first presented by Lavallée (1995) in the context of the problem of cross-sectional weighting for longitudinal household surveys. Since then, the author produced new theoretical and practical results on the GWSM. Recently, some survey statisticians drew on these results in order to solve concrete problems associated with the indirect sampling of clusters. We present here six of these new applications of the GWSM that were chosen not necessarily for their complexity, but instead for their diversity.

A first application is the one from Whitridge and Beaucage (2000) in relation to the measurement by Statistics Canada of the extent of electronic commerce in Canada. Here, we want to produce estimates on the use of electronic commerce for which the variables of interest y related to the use of the computer, the Internet and Web pages are measured. For example, we are interested in the proportion of enterprises that perform transactions over the Internet. This study covers all economic sectors with the exception of local governments, some agricultural sectors, construction and fishing. Note that the use of electronic commerce is a phenomenon still considered as relatively rare in Canada. The target population is the universe of enterprises from the economic sectors mentioned earlier.

An important constraint associated with this study is the use of the sample selected for the Canadian Public and Private Investment Survey (Statistics Canada, 2000b). This sample is one of establishments that covers the same economic sectors as the target population. This constraint was imposed in order to reduce as much as possible the selection, contact, and collection costs related to the study. Note that the selection of establishments instead of enterprises allows the sample selection to be controlled at the geographic and sector levels. The sampling frame U^A is therefore the universe of establishments covering the same economic sectors as the target population U^B. Thus, a sample s^A of establishments selected from U^A is used to survey the enterprises having the establishments in s^A. This situation is illustrated in Figure 3.2. The GWSM here offers a simple solution to produce estimates and their variance.

A second application is the one described by Girard and Simard (2000) in the context of Statistics Canada's Unified Enterprise Survey (UES). This survey is part of an extensive project known under the acronym PIPES (Project to Improve Provincial Economic Statistics) that has an objective of implementing a complete system of annual economic statistics by province, from business and household statistics, as well as from data drawn from tax records and other sources of administrative data. Important secondary objectives are also intended: the reduction of response burden for Canadian businesses; the use of a single sampling frame, the Business Register; and the development of an approach and integrated methods that are the most coherent possible for all annual economic programs at Statistics Canada (Laniel and Royce, 1998).

PIPES is based on the use of network sampling described in section 3.4. This type of sampling is used here to select enterprises through their establishments. Hence, a sample of establishments is selected and subsequently, the enterprises having these selected establishments are surveyed. The establishments are sampled instead of the enterprises because we want to control the sample selection at the geographic and sector levels. The population U^A is therefore the universe of establishments that covers the majority of the economic sectors. The target population U^B is the population of enterprises covering the same economic sectors as U^A. Note that this population consists of clusters of size 1. Girard and Simard (2000) considered two options for the production of estimates from UES and the calculation of their variance. The first option is based on the exact calculation of the selection probabilities (say π_{ik}^B) of the enterprises ik surveyed through the sample s^A, while the second option depends on the use of the GWSM. It turns out that the GWSM offers a much simpler solution than the exact calculation of the probabilities, particularly with regard to the calculation of the variances.

A third application of the GWSM is the one from Ardilly and Le Blanc (2001) that used the GWSM to weight a survey of homeless persons. The problem with this type of survey is the absence of a sampling frame for the target population U^B, which is here the set of homeless persons in France. Thus, an indirect sampling is required. The variables of interest of this survey are, for example, the age at the end of the studies and the number of centres frequented. To survey these persons, Ardilly and Le Blanc (1999) and Ardilly and Le Blanc (2001) proposed to make use of the services provided to these persons in the centres during a certain reference period such as a day, a week, or a month. A service can be, for example, a meal or an accommodation. The population U^A from which the sample is selected is therefore the set of services provided during the chosen reference period. Each service from U^A is linked to a homeless person from U^B and, of course, a homeless person can receive more than one service. Thus, we are in the situation of "many-to-one" links between the populations U^A and U^B, as illustrated in Figure 3.2. Ardilly and Le Blanc (2001) used the GWSM to produce estimates on the homeless persons. Although they were confronted with a problem of identification of links, the GWSM again proved to be greatly useful.

A fourth application of indirect sampling is the one from Deville and Maumy (2005) where indirect sampling was used to measure tourism in the region of Brittany in France. This application is somewhat similar to the one from Ardilly and Le Blanc (2001), as we try to measure a target population of people (in the present example, tourists) using a frame based on services provided to them. The application of Deville and Maumy (2005) differs however from the fact that the frame used by the survey has been built from three different frames: (1) a subset of the most visited attractions in Brittany; (2) the highway payment poll of La Gravelle that most automobiles used to enter or leave Brittany; (3) a sample of bakeries. Note that these services have been sampled for a given time period.

For each of the three different frames (or populations) U_q^A, a sample s_q^A of tourists has been selected to estimate the total population U^B of tourists in Brittany. Now, it is clear that a given tourist can be found in all three frames, since he is likely to visit the main attractions of Brittany, use the highway, or buy some bread. Therefore, we are in a context of "many-to-one" links between the populations U^A and U^B. Using the information collected from the three samples s_q^A, estimates have been produced for the target population U^B using weights obtained through the GWSM.

It should be noted that the GWSM is offering here a different way to attack the estimation problem in the context of *multiple frames*. This problem has been known for years, and the related theory has been developed by Hartley (1962). For more details on multiple frame estimation, one can see Kott and Vogel (1995).

A fifth application of indirect sampling is the one from Dessertaine and Fluteaux (2004), which is in the context of traditional mailing in France. The problem was to estimate the flow of mail at *La Poste*, the French national mail agency. The population U^A is rounds of postmen j, while the target population U^B is objects k (envelopes, packages, etc.) distributed at a given day i (cluster). The links $l_{j,ik}$ between the two populations relate the postman j to the objects k that he delivered on a given day i. Note that because it was difficult to establish exactly how many objects were delivered by a certain postman k on a given day i, Dessertaine and Fluteaux (2004) were faced with the problem of obtaining the total number of links $L_{j,i}$, which is a problem of links identification. They solved this problem by using, instead of the links $l_{j,ik}$, the probability $\theta_{j,ik}$ of having a link. They obtained a mathematical formulation of the estimator of the total Y^B similar to (9.7).

Finally, Renaud (2006) used the GWSM to weight the sample of towns for the estimation of the 2004 Swiss statistics on social security beneficiaries. The sample of towns was selected in 1999 from a list of towns established in 1998. From 1998 to 2004, some modifications occurred to the towns, and therefore the weights needed to be adjusted to account for these changes. This situation is similar to the one illustrated in Figure 1.4. As some towns were collapsed to others, or divided into smaller towns, the links between the population of towns in 1998 (population U^A) and the one in 2004 (population U^B) were complex. The use of the GWSM turned out to be useful to solve this estimation problem.

In the future, we expect other developments around the GWSM. For example, we can think of the development of allocation methods for the sample s^A, considering that we are faced with an indirect sampling. These methods could consider cost constraints, in addition to constraints in precision.

In closing, the author knows that the developments presented in this book only represent the tip of the iceberg of the potential of indirect sampling and the GWSM. The more indirect sampling is studied, the more its potential to solve, in a simple manner, complex estimation problems is discovered. The GWSM opens up new possibilities to simply treat theoretical and practical situations that are introduced during the use of sample surveys to obtain information.

NOTATIONS

α	Constant used in the definition of the weight share method
β	Regression coefficient
$\boldsymbol{\beta}$	Column vector of regression coefficients
δ	Indicator variable
∇	Gradient function
π	Probability of selection
$\boldsymbol{\Pi}$	Diagonal matrix of selection probabilities
Ω	Set of surveyed clusters
$\boldsymbol{\Xi}$	Set of all samples s
θ	Variable identifying the weighted links
$\boldsymbol{\Theta}$	Matrix of variables identifying the weighted links
$\boldsymbol{\gamma}$	Column vector of auxiliary variables
$\boldsymbol{\Gamma}$	Column vector containing the total of the auxiliary variables $\boldsymbol{\gamma}$
υ	Indicator variable indicating a match between two records
Δ	Indicator variable associated with the comparison of two records
$\boldsymbol{\Delta}$	Matrix entering into the expression of the variance
ς	Subscript identifying the comparisons
η	Number of clusters (households) from U^A
ι	Clusters (household) from the population U^A
κ	Number of names from snowball sampling
τ	Number of phases from snowball sampling
μ	Cluster average in adaptive cluster sampling
σ^2	Variance
ω	Multiplicity weight related with network sampling
$\boldsymbol{\lambda}$	Lagrange multiplier
Λ	Likelihood function
ϕ	Probability of response for a unit
Φ	Probability of response for a cluster of units

$\boldsymbol{\psi}$	Column vector derived from auxiliary variables $\mathbf{x}$ and the variable of interest y
$\boldsymbol{\Psi}$	Matrix derived from auxiliary variables $\mathbf{x}$
ν	Random number uniformly distributed between]0,1[
ϑ	Statistic
$\mathbf{1}$	Column vector of 1's
A	Superscript identifying the population for which we have a sampling frame
$ADAP$	Superscript identifying adaptive cluster sampling
B	Superscript associated with the target population
c	Random (or repeated) group
C	Number of random (or repeated) groups
CAL	Superscript identifying calibration
$CALG$	Superscript identifying generalized calibration
$CLUS$	Superscript identifying cluster sampling
$COND$	Superscript identifying the conditional approach to improve estimators
D	Decision rule of Fellegi and Sunter
$\mathbf{D}$	Set of indices and measured variables for a sample
e	Regression residual
f	Sampling fraction
F	Inverse of the derivative of the distance function G
FS	Superscript identifying the approach of Fellegi and Sunter
g	Derivative of the distance function G
G	Superscript identifying the intermediate population obtained through factorisation, and distance function used in calibration
$GLOB$	Superscript identifying the global approach for unit non-response
$GPROP$	Superscript identifying the use of global proportional adjustment
h	Stratum
$\mathbf{h}$	Function
HT	Superscript identifying the Horvitz-Thompson estimator
i	Cluster from the population U^B
I	Interval
$\mathbf{I}$	Identity matrix

II	Superscript identifying two-stage indirect sampling
j	Unit from the population U^A
$JACK$	Superscript identifying the use of the Jackknife method
k	Unit from the population U^B
l	Indicator variable identifying the links between U^A and U^B
L	Total number of links
$\mathbf{L}$	Set of all links
$LGLIN$	Superscript identifying the use of the log-linear model
$LOGIT$	Superscript identifying the use of the logistic model
m	Number of units selected in the sample
M	Number of units from the population
$MULT$	Superscript identifying the multiplicity approach
n	Number of surveyed clusters
N	Number of clusters from the target population
NET	Superscript identifying network sampling
NR	Superscript identifying non-response
NRA	Superscript identifying the case of non-response within s^A
NRC	Superscript identifying the case of non-response of clusters
NRL	Superscript identifying the problem of links identification
NRU	Superscript identifying the case of non-response of units
opt	Superscript identifying an optimal quantity
$\mathbf{p}$	Sampling plan
p	Dimension of the vectors of auxiliary variables
$PROP$	Superscript identifying the use of proportional adjustment
q	Group or subset from the population
Q	Total number of groups or subsets from the population
r	Subcript identifying the subset of respondents
R	Corrected response rate
RB	Superscript identifying the use of the Rao-Blackwell theorem
REG	Superscript identifying the regression estimator
RL	Superscript identifying record linkage
RLT	Superscript identifying record linkage with threshold
s	Sample
$SUBS$	Superscript identifying the use of subsampling

t	Indicator variable identifying the units selected in U^A
T	Transpose of a matrix or vector
$\mathbf{T}$	Diagonal matrix of indicator variables t
u	Sufficient statistic
U	Population
w	Estimation weight
$\mathbf{W}$	Column vector of estimation weights
WSM	Superscript identifying the weight share method
$\mathbf{x}$	Column vector of auxiliary variables
$\mathbf{X}$	Column vector containing the total of the auxiliary variables $\mathbf{x}$
y	Variable of interest
Y	Total of the variable of interest y
$\mathbf{Y}$	Column vector containing the variable of interest y
z	Variable derived from the variable of interest y
Z	Total of the derived variable z
$\mathbf{Z}$	Column vector of variables derived from y

BIBLIOGRAPHY

ARDILLY, P. (2006). *Les techniques de sondage, 2ème édition*. Éditions Technip, Paris, 696 pages.

ARDILLY, P., LE BLANC, D. (1999). *Enquête auprès des personnes sans domicile : éléments techniques sur l'échantillonnage et le calcul de pondérations individuelles – Une application de la méthode du partage des poids*. INSEE working document, No. F9903.

ARDILLY, P., LE BLANC, P. (2001). Comment pondérer une enquête auprès des personnes sans domicile?. *Enquêtes, modèles et applications*, Dunod, Paris, 417-436.

BANKIER, M. (1983). *Evaluation of the Partnership Correction for TRA's T1 Weights*. Statistics Canada internal document, January 13 1983.

BARTLETT, S., KREWSKI, D., WANG, Y., ZIELINSKI, J.M. (1993). Evaluation of Error Rates in Large Scale Computerized Record Linkage Studies. *Survey Methodology*, Vol. 19, No. 1, pp. 3-12.

BASU, D. (1958). On Sampling With and Without Replacement. *Sankhyã*, Vol. 20, pp. 287-294.

BELIN, T.R. (1993). Evaluation of Sources of Variation in Record Linkage through a Factorial Experiment. *Survey Methodology*, Vol. 19, No. 1, pp. 13-29.

BERNIER, N., LAVALLÉE, P. (1994). *La macro SAS : CALJACK, version 2.04*. Statistics Canada internal document, December 1994.

BICKEL, P.J., DOKSUM, K.A. (1977). *Mathematical Statistics: Basic Ideas and Selected Topics*. Holden-Day Inc., Oakland, California, 493 pages.

BISHOP, Y.M.M., FINBERG, S.E., HOLLAND, P.W. (1975). *Discrete Multivariate Analysis, Theory and Pratice*. MIT Press, Cambridge, Massachusetts, 1975.

BIRNBAUM, Z.W., SIRKEN, M.G. (1965). Design of Sample Surveys to Estimate the Prevalence of Rare Diseases: Three Unbiased Estimates. *Vital and Health Statistics,* PHS Publication No. 1000-Series 2, No. 11, National Center for Health Statistics, Washington, D.C., October 1965.

BLACKWELL, D. (1947). Conditional Expectation and Unbiased Sequential Estimation. *Annals of Mathematical Statistics,* Vol. 18, pp. 105-110.

BOGESTRÖM, B., LARSSON, M., LYBERG, L. (1983). Bibliography on Nonresponse and Related Topics. In *Incomplete Data in Sample Surveys* (Eds. W.G. Madow, I. Olkin and D.B. Rubin), Vol. 2, Academic Press, New York, pp. 479-567.

BUDD, E.C. (1971). The creation of a microdata file for estimating the size distribution of income. *The Review of Income and Wealth,* Vol. 17, pp. 317-333.

BUDD, E.C., RADNER, D.B. (1969). The OBE size distributions series: methods and tentative results for 1964. *American Economic Review, Papers and Proceedings*, LIX, pp. 435-449.

CASSEL, C.-M., SÄRNDAL, C.-E., WRETMAN, J.H. (1976). Some Results on Generalized Difference Estimation and Generalized Regression Estimation for Finite Populations. *Biometrika*, Vol. 63, pp. 615-620.

CASSEL, C.-M., SÄRNDAL, C.-E., WRETMAN, J.H. (1977). *Foundations of Inference in Survey Sampling.* John Wiley and Sons, New York, 192 pages.

COCHRAN, W.G. (1977). *Sampling Techniques, third edition.* John Wiley and Sons, New York, 428 pages.

COLLETT, D. (1991). *Modelling Binary Data.* Chapman and Hall, New York, 369 pages.

DESSERTAINE, A., FLUTEAUX, L. (2004). Utilisation de la Méthode généralisée du partage des poids dans le cadre des estimations de flux de courrier à La Poste. In *Échantillonnage et méthodes d'enquêtes* (Ed. P. Ardilly), Dunod, Paris.

DEVILLE, J.-C. (1988). Estimation linéaire et redressement sur information auxiliaires d'enquêtes par sondage. In *Essais en l'honneur d'Edmond Malinvaud* (Eds. A. Monfort and J.J. Laffont), Economica, Paris, pp. 915-927.

DEVILLE, J.-C. (1998a). Les enquêtes par panel : en quoi diffèrent-elles des autres enquêtes ? suivi de : comment attraper une population en se servant d'une autre. *INSEE Méthodes*, No. 84-85-86, pp. 63-82.

DEVILLE, J.-C. (1998b). La correction de la non-réponse par calage ou par échantillonnage équilibré. *1998 Proceedings of the Survey Methods Section*, Statistical Society of Canada, pp. 103-110.

DEVILLE, J.-C. (2000a). Simultaneous Calibration of Several Surveys. *Proceedings of Statistics Canada Symposium 99: Combining Data from Different Sources*, Publication No. 11-522-XCB, Statistics Canada, Ottawa, September 2000, pp. 225-230.

DEVILLE, J.-C. (2000b). Generalized calibration and application to weighting for non-response. *Proceedings in Computational Statistics 2000* (Eds. J.G. Bethlehem and P.G.M. van der Heijden), Physica-Verlag, New York, pp. 65-76.

DEVILLE, J..-C., MAUMY, M. (2005). Extension de la méthode d'échantuillonnage indirect et son application dans l'enquête sur le tourisme: ARGOAT. *Actes des Journées de Méthodologie Statistique de l'INSEE*, Paris, February 2005.

DEVILLE, J.-C., SÄRNDAL, C.-E. (1992). Calibration Estimators in Survey Sampling. *Journal of the American Statistical Association*, Vol. 87, No. 418, June 1992, pp. 376-382.

DROESBEKE, J.-J., LAVALLÉE, P. (1996). La non-réponse dans les enquêtes. *Methodologica,* No. 4, pp. 1-39.

DUFOUR, J., GAMBINO, J. KENNEDY, B. LINDEYER, J. SINGH, M.P. (1998). *Methodology of the Canadian Labour Force Survey.* Statistics Canada, Catalogue 71-526-XPB.

ERNST, L. (1989). Weighting issues for longitudinal household and family estimates. In *Panel Surveys* (Eds. D. Kasprzyk, G. Duncan, G. Kalton and M.P. Singh), John Wiley and Sons, New York, pp. 135-159.

ERNST, L., HUBBLE, L., JUDKINS, D.R. (1984). Longitudinal Family and Household Estimation in SIPP. Survey Research Section of the *Proceedings of the American Statistical Association*, pp. 682-687.

FELLEGI, I.P., SUNTER, A. (1969). A theory for record linkage. *Journal of the American Statistical Association*, Vol. 64, pp. 1183-1210.

FULLER, W.A., ISAKI, C.T. (1981). Survey Design Under Superpopulation Models. In *Current Topics in Survey Sampling* (Eds. D. Krewski, J.N.K. Rao and R. Platek), Academic Press, New York, pp. 199-226.

GAILLY, B., LAVALLÉE, P. (1993). Insérer des nouveaux membres dans un panel longitudinal de ménages et d'individus : simulations. *CEPS/Instead, Document PSELL,* No. 54, Luxembourg, May 1993.

GIRARD, C., SIMARD, M. (2000). *Network Sampling: Actual application in a major survey.* Article presented at the International Conference on Establishment Surveys – II, Buffalo, June 2000.

GOODMAN, L.A. (1961). Snowball Sampling. *Annals of Mathematical Statistics*, Vol. 32, No. 1, pp. 148-170.

GRANOVETTER, M. (1976). Network Sampling: Some First Steps. *American Journal of Sociology*, Vol. 81, No. 6, pp. 1287-1303.

GROSBRAS, J.-M. (1986). *Méthodes statistiques des sondages.* Economica, Paris.

HARTLEY, H.O. (1962). Multiple Frame Surveys. *Proceedings of the Social Statistics Section*, American Statistical Association, pp. 203-206.

HEDGES, L.V., OLKIN, I. (1983). Selected Annotated Bibliography. In *Incomplete Data in Sample Surveys* (Eds. W.G. Madow, I. Olkin and D.B. Rubin), Vol. 2, Academic Press, New York, pp. 417-478.

HORVITZ, D.G., THOMPSON, D.J. (1952). A Generalization of Sampling Without Replacement from a Finite Universe. *Journal of the American Statistical Association*, Vol. 47, pp 663-685.

HUANG, H. (1984). Obtaining Cross-Sectional Estimates from a Longitudinal Survey: Experiences of the Income Survey Development Program. Survey Research Section of the *Proceedings of the American Statistical Association*, 1984, pp. 670-675.

HURAND, C. (2006). *La Méthode généralisée du partage des poids et le problème d'identification des liens.* Work term report, Université de Lyon 2, Lyon, France, July 2006.

ISAKI, C.T., FULLER, W.A. (1982). Survey Designs Under the Regression Superpopulation Model. *Journal of the American Statistical Association*, Vol. 77, pp. 89-96.

JAZMINSKI, A.H. (1970). *Stochastic Processes and Filtering Theory.* Academic Press, New York.

JUDKINS, D., HUBBLE, D., DORSCH, J., MCMILLEN, D., ERNST, L. (1984). Weighting of Persons for SIPP Longitudinal Tabulations. Survey Research Section of the *Proceedings of the American Statistical Association*, pp. 676-681.

KALTON, G., BRICK, J.M. (1995). Weighting Schemes for Household Panel Surveys. *Survey Methodology*, Vol. 21, No. 1, pp. 33-44.

KOTT, PH.S., VOGEL, F.A. (1995). Multiple-Frame Business Surveys. In *Business Survey Methods* (Eds. B.A. Cox, D.A. Binder, B.N. Chinnappa, A. Christianson, M.J. Colledge and Ph.S Kott), John Wiley and Sons, New York. pp.185-203.

LANIEL, N., ROYCE, D., (1998). Projet d'amélioration des statistiques économiques provinciales : Objectifs et Enquêtes-pilotes. *1998 Proceedings of the Survey Methods Section*, Statistical Society of Canada, pp. 59-63.

LAVALLÉE, P. (1993). Sample representativity for the survey of labour and income dynamics. *Survey of Labour and Income Dynamics working document*, Statistics Canada, Catalogue No. 93-19, December 1993.

LAVALLÉE, P. (1995). Cross-sectional Weighting of Longitudinal Surveys of Individuals and Households Using the Weight Share Method. *Survey Methodology*, Vol. 21, No. 1, pp. 25-32.

LAVALLÉE, P. (1998a). Business Panel Surveys: Following enterprises versus following establishments. *Research in Official Statistics*, Eurostat, Vol. 0, pp. 37-57.

LAVALLÉE, P. (1998b). *Théorie et application des enquêtes longitudinales.* Course 411F notes offered at Statistics Canada, Ottawa, October 1998.

LAVALLÉE, P., CARON, P. (2001). Estimation Using the Generalised Weight Share Methods: The Case of Record Linkage. *Survey Methodology*, Vol. 27, No. 2, pp. 155-169.

LAVALLÉE, P., DEVILLE, J.-C. (2002). Theoretical Foundations of the Generalised Weight Share Method. *Proceedings of the International*

Conference on Recent Advances in Survey Sampling, Laboratory for Research in Statistics and Probability (Technical Report No. 386), Ottawa, 127-136.

LAVALLÉE, P., HUNTER, L. (1993). Weighting for the Survey of Labour and Income Dynamics. *Proceedings of Statistics Canada Symposium 92: Design and Analysis of Longitudinal Surveys*, Publication No. 11-522-XPE, Statistics Canada, Ottawa, August 1993, pp. 65-75.

LAVIGNE, M., MICHAUD, S. (1998). General Aspects of the Survey of Labour Income Dynamics. *Survey of Labour and Income Dynamics working document*, Statistics Canada, Catalogue No. 98-05, March 1998.

LE GUENNEC, J., SAUTORY, O. (2004). CALMAR 2: Une nouvelle version de la macro Calmar de redressement d'échantillon par calage. In *Échantillonnage et méthodes d'enquêtes*, Dunod, 375 pages.

LEMAÎTRE, G., DUFOUR, J. (1987). An Integrated Method for Weighting Persons and Families. *Survey Methodology*, Vol. 13, No. 2, pp. 199-207, December 1987.

LEMEL, Y. (1976). Une généralisation de la méthode du quotient pour le redressement des enquêtes par sondage. *Annales de l'INSEE*, Vol. 22-23, pp. 272-282.

LEPKOWSKI, J.M. (1989). Treatment of Wave Nonresponse in Panel Surveys. In *Panel Surveys* (Eds. D. Kasprzyk, G. Duncan, G. Kalton and M.P. Singh, M.P.), John Wiley and Sons, New York, pp. 348-374.

LÉVESQUE, I., FRANKLIN, S. (2000). *Longitudinal and Cross-sectional Weighting of the Survey of Labour and Income Dynamics, 1997 Reference Year.* Statistics Canada, Catalogue No. 75F0002MIF-00004, August 2000.

LEVY, P.S. (1977). Optimum Allocation in Stratified Random network Sampling for Estimation the Prevalence of Attributes in Rare Populations. *Journal of the American Statistical Association*, Vol. 72, No. 360, pp. 758-763.

LIM, A. (2000). *Results of the Linkage between the 1998 Taxation Data and the 1998 Farm Register*. Business Survey Methods Division internal document, Statistics Canada, July 7 2000.

LOCK OH, H., SCHEUREN, F.J. (1983). Weighting Adjustment for Unit Nonresponse. In *Incomplete Data in Sample Surveys* (Eds. W.G. Madow, I. Olkin and D.B.), Vol. 2, Academic Press, New York, pp. 143-184.

LOHR, S. (1999). *Sampling: Design and Analysis*. Duxbury Press, California, 494 pages.

MICHAUD, S., HUNTER, L. (1992). Strategy for Minimizing the Impact of Nonresponse for the Survey of Labour and Income Dynamics. *Proceedings of Statistics Canada Symposium 92: Design and Analysis of*

Longitudinal Surveys, Publication No. 11-522-XPE, Statistics Canada, Ottawa, August 1993, pp. 89-98.

MORIN, H. (1993). *Théorie de l'échantillonnage.* Presses de l'Université Laval, Ste-Foy, 178 pages.

NEWCOME, H.B., KENNEDY, J.M., AXFORD, S.J., JAMES, A.P. (1959), Automatic Linkage of Vital Records. *Science*, Vol. 130, pp. 954-959.

OKNER, B.A. (1972). Constructing a new data base from existing microdata sets: The 1966 merge file. *Annals of Economic and Social Measurement,* Vol. 1, pp. 325-342.

PFEFFERMANN, D. (1993). The Role of Sampling Weights when Modeling Survey Data. *International Statistical Review*, Vol. 61, No. 2, pp. 317-337.

PLATEK, R., GRAY, G.B. (1983). Imputation Methodology. In *Incomplete Data in Sample Surveys* (Eds. W.G. Madow, I. Olkin and D.B. Rubin), Vol. 2, Academic Press, New York, pp. 255-294.

RAO, C.R. (1945). Information and Accuracy Attainable in Estimation of Statistical Parameters. *Bulletin of the Calcutta Mathematical Society,* Vol. 37, pp. 81-91.

RENAUD, A. (2006). Statistique suisse des bénéficiaires de l'aide social, Pondération des communes 2004. *Rapport de méthodes*, Office fédérale de la statistique, Neuchâtel, Switzerland.

RUBIN, D. B. (1976). Inference and Missing Data. *Biometrika*, Vol. 63, pp. 581-592.

RUBIN, D. B. (1983). Conceptual issues in the presence of nonresponse. In *Incomplete Data in Sample Surveys* (Eds. W.G. Madow, I. Olkin and D.B. Rubin), Vol. 2, Academic Press, New York, pp. 123-142.

RUBIN, D. B. (1987). *Multiple Imputation for Nonresponse in Surveys.* John Wiley and Sons, New York.

SANDERS, L.L., KALSBEEK, W.D. (1990). Network Sampling as an Approach to Sampling Pregnant Women. Survey Research Section of the *Proceedings of the American Statistical Association*, pp. 326-331.

SÄRNDAL, C.-E., SWENSSON, B., WRETMAN, J. (1992). *Model Assisted Survey Sampling.* Springer-Verlag, New York.

SAUTORY, O. (1991). *La macro SAS : CALMAR (redressement d'un échantillon par calage sur marges).* INSEE internal document, Paris.

SAUTORY, O. (1992). Redressement d'échantillons d'enquêtes auprès des ménages par calage sur marges. Actes des journées de méthodologie statistique, March 13-14 1991, *INSEE Méthodes*, No. 29-30-31, December 1992, pp. 299-326.

SAUTORY, O. (1993). *Méthodes de pondération des ménages et des individus dans les enquêtes.* Document presented at the "XXVes Journées de statistique" of the Association pour la Statistique et ses Utilisations (ASU), Vannes, May 24-28 1993.

SEARLE, S.R. (1971). *Linear Models*. John Wiley and Sons, New York, 532 pages.

SINGH, A.C., MANTEL, A.J., KINACK, M.D., ROWE, G. (1993). Statistical Matching: Use of Auxiliary Information as an Alternative to the Conditional Independence Assumption. *Survey Methodology*, Vol. 19, No. 1, pp. 59-79.

SINGH, M.P., DREW, J.D., GAMBINO, J.G., MAYDA, F. (1990). *Methodology of the Canadian Labour Force Survey.* Statistics Canada, Catalogue 71-526.

SIRKEN, M.G. (1970). Household Surveys with Multiplicity. *Journal of the American Statistical Association*, Vol. 65, No. 329, pp. 257-266.

SIRKEN, M.G. (1972). Stratified Sample Surveys with Multiplicity. *Journal of the American Statistical Association*, Vol. 67, No. 337, pp. 224-227.

SIRKEN, M.G., LEVY, P.S. (1974). Multiplicity Estimation of Proportions Based on Ratio of Random Variables. *Journal of the American Statistical Association*, Vol. 69, No. 345, pp. 68-73.

SIRKEN, M.G., NATHAN, G. (1988). Hybrid Network Sampling. Survey Research Section of the *Proceedings of the American Statistical Association*, pp. 459-461.

SIRKEN, M.G., SHIMIZU, I. (1999). Population Based Establishment Sample Surveys: The Horvitz-Thompson Estimator. *Survey Methodology*, Vol. 25, No. 2, pp. 187-191.

STATISTICS CANADA (2000a). *Whole Farm Data Base, Reference Manual.* Publication No. 21F0005GIE, Statistics Canada, Ottawa, January 2000, 100 pages.

STATISTICS CANADA (2000b). *Public and Private Investment in Canada, Intentions 2000.* Publication No. 61-205-XIB, Statistics Canada, Ottawa, February 2000, 155 pages.

THOMPSON, S.K. (1990). Adaptive Cluster Sampling. *Journal of the American Statistical Association*, Vol. 85, No. 412, pp. 1050-1059.

THOMPSON, S.K. (1991a). Stratified Adaptive Cluster Sampling. *Biometrika*, Vol. 78, No. 2, pp. 389-397.

THOMPSON, S.K. (1991b). Adaptive Cluster Sampling: Designs with Primary and Secondary Units. *Biometrics*, Vol. 47, September 1991, pp. 1103-1115.

THOMPSON, S.K. (1992). *Sampling.* John Wiley and Sons, New York.

THOMPSON, S.K. (2002). *Sampling, 2nd Edition.* John Wiley and Sons, New York, 400 pages.

THOMPSON, S.K., SEBER, G.A. (1996). *Adaptive Sampling.* John Wiley and Sons, New York.

WHITRIDGE, P., BEAUCAGE, Y. (2000). *Statistics Canada's Electronic Commerce Survey.* Business Survey Methods Division internal document, Statistics Canada, October 6 2000.

WINKLER, W.E. (1995). Matching and Record Linkage. In *Business Survey Methods* (Eds. B.G. Cox, D.A. Binder, B.N. Chinnappa, A. Christianson, M.J. Colledge and P. Kott), John Wiley and Sons, New York, pp. 355-384.

WOLTER, K.M. (1985). *Introduction to Variance Estimation.* Springer-Verlag, New York.

XU, X., LAVALLÉE, P. (2006). *Treatment of Links Nonresponse in Indirect Sampling.* Paper presented at the annual conference of the Statistical Society of Canada, London, Ontario, June 2006.

YATES, F., GRUNDY, P.M. (1953). Selection Without Replacement from Within Strata with Probability Proportional to Size. *Journal of the Royal Statistical Society B*, Vol. 15, pp. 235-261.

INDEX

A
Adaptive cluster sampling, 37
Attrition, 152

B
Bernoulli distribution, 155, 165, 173
Bernoulli trials, 204
Bias, 3, 45

C
Calibration, 119
Calibration weight, 119
CALJACK, 123
CALMAR, 121
Census, 53
Classical approach, 197
Cluster effect, 156
Cluster non-response, 151
Cluster sampling, 4, 50
Clusters, 4
Coefficient of variation, 209
Cohabitants, 107
Complex links, 20
Conditional approach, 57
Corrective response rate, 153
Cross-sectional estimates, 103

D
Distance function, 120

E
Edge units, 37
Enumeration units, 32
Equal household weighting scheme, 31
Equal person weighting scheme, 32
Estimation weight, 4, 15
Euclidian distance, 121
Exact matching, 194

F
Factorisation, 67
Fair share method, 24
Final weight, 16
Fixed size sampling, 1

G
Generalised calibration, 124
Generalised inverse, 122
Generalised regression estimator, 121
Generalised weight share method, 10
Global approach, 176
Global proportional adjustment, 190
g-Weight, 121

H
Horvitz-Thompson estimator, 2

I

Ignorable non-response, 152
Inclusion probability, 2
Indirect sampling, 7
Individual approach, 176
Initial weight, 16
Initially-absent individuals, 106
Initially-present individuals, 106
Instrumental variable, 124
Integrated calibration, 140
Integrated weighting, 140
Intracorrelation, 156
Immigrants, 106

J

Jackknife method, 115

L

Link, 13
Link matrix, 64
Linkage weight, 195
Log-linear model, 186
Logistic regression model, 157
"Logit" model, 157
Longitudinal households, 107
Longitudinal individuals, 106
Longitudinal surveys, 103

M

Multiple frames, 226
Multiplicity approach, 24, 27
Multiplicity estimation, 34
Multiplicity weight, 35

N

Network sampling, 32
Network, 32, 38
Non-response, 149
NYSIIS, 210

O

Operational response rate, 153
Optimal link matrix, 68
Optimal weighted links, 63

P

Partial non-response, 150
Primary sampling unit, 4
Probabilistic linkage, 193
Probability sampling, 2
Problem of links identification, 151
Proportional adjustment, 189
Proxy, 151

R

Random size sampling, 1
Rao-Blackwell theorem, 61
Rare populations, 19
Record linkage, 185, 194
Relationship, 13
Response homogeneity groups, 157
Response probability, 152

S

Sampling design, 2
Sampling fraction, 146, 212
Sampling frame, 1
Sampling weight, 3
Secondary sampling unit, 4
Selection probability, 2
Simple random sampling, 2
Snowball sampling, 42
Standardised link matrix, 65

Statistical matching, 194
Strata, 2
Stratified simple random sampling, 2
Strong optimality, 70
Sufficient statistics, 61
Supplementary sample, 107

T

Target population, 1
Total non-response, 150
Two-stage indirect sampling, 76
Two-stage sampling, 4

U

Units, 1
Unit non-response, 151

V

Variance, 3, 45

W

Wave non-response, 152
Weak optimality, 70
Weight share method, 28
Weighted links, 55

Springer Series in Statistics *(continued from p. ii)*

Küchler/Sørensen: Exponential Families of Stochastic Processes.
Kutoyants: Statistical Inference for Ergodic Diffusion Processes.
Lahiri: Resampling Methods for Dependent Data.
Lavallée: Indirect Sampling.
Le Cam: Asymptotic Methods in Statistical Decision Theory.
Le Cam/Yang: Asymptotics in Statistics: Some Basic Concepts, 2nd edition.
Le/Zidek: Statistical Analysis of Environmental Space-Time Processes.
Liu: Monte Carlo Strategies in Scientific Computing.
Manski: Partial Identification of Probability Distributions.
Mielke/Berry: Permutation Methods: A Distance Function Approach, 2nd edition.
Molenberghs/Verbeke: Models for Discrete Longitudinal Data.
Mukerjee/Wu: A Modern Theory of Factorial Designs.
Nelsen: An Introduction to Copulas, 2nd edition.
Pan/Fang: Growth Curve Models and Statistical Diagnostics.
Politis/Romano/Wolf: Subsampling.
Ramsay/Silverman: Applied Functional Data Analysis: Methods and Case Studies.
Ramsay/Silverman: Functional Data Analysis, 2nd edition.
Reinsel: Elements of Multivariate Time Series Analysis, 2nd edition.
Rosenbaum: Observational Studies, 2nd edition.
Rosenblatt: Gaussian and Non-Gaussian Linear Time Series and Random Fields.
Särndal/Swensson/Wretman: Model Assisted Survey Sampling.
Santner/Williams/Notz: The Design and Analysis of Computer Experiments.
Schervish: Theory of Statistics.
Shao/Tu: The Jackknife and Bootstrap.
Simonoff: Smoothing Methods in Statistics.
Sprott: Statistical Inference in Science.
Stein: Interpolation of Spatial Data: Some Theory for Kriging.
Taniguchi/Kakizawa: Asymptotic Theory for Statistical Inference for Time Series.
Tanner: Tools for Statistical Inference: Methods for the Exploration of Posterior Distributions and Likelihood Functions, 3rd edition.
Titlé: Sampling Algorithms.
Tsaitis: Semiparametric Theory and Missing Data.
van der Laan/Robins: Unified Methods for Censored Longitudinal Data and Causality.
van der Vaart/Wellner: Weak Convergence and Empirical Processes: With Applications to Statistics.
Verbeke/Molenberghs: Linear Mixed Models for Longitudinal Data.
Weerahandi: Exact Statistical Methods for Data Analysis.

MIX
Papier aus verantwortungsvollen Quellen
Paper from responsible sources
FSC® C105338

If you have any concerns about our products,
you can contact us on
ProductSafety@springernature.com

In case Publisher is established outside the EU,
the EU authorized representative is:
Springer Nature Customer Service Center GmbH
Europaplatz 3, 69115 Heidelberg, Germany

Printed by Libri Plureos GmbH
in Hamburg, Germany